Student Support Materials for AQA

A2 Biology

Unit 4: Populations and Environment

Author: Mike Boyle

Series Editors: Keith Hirst and Lesley Higginbottom

William Collins's dream of knowledge for all began with the publication of his first book in 1819. A self-educated mill worker, he not only enriched millions of lives, but also founded a flourishing publishing house. Today, staying true to this spirit, Collins books are packed with inspiration, innovation and practical expertise. They place you at the centre of a world of possibility and give you exactly what you need to explore it.

Collins. Freedom to teach.

Published by Collins
An imprint of HarperCollinsPublishers
77-85 Fulham Palace Road
Hammersmith
London
W6 8JB

Browse the complete Collins catalogue at
www.collinseducation.com

10 9 8 7 6 5 4 3 2 1

ISBN-13 978-0-00-726819-1

British Library Cataloguing in Publication Data. A Catalogue record for this publication is available from the British Library.

Commissioned by Penny Fowler
Series Editors Keith Hirst and Lesley Higginbottom
Project Managed by Alexandra Riley
Edited by Camilla Behrens
Proof read by Rachel Hutchings
Design by Newgen Imaging
Cover design by Angela English
Index by Laurence Errington
Production by Arjen Jansen
Printed and bound in Hong Kong by Printing Express

Contents

Populations and environment

Topics at a glance. This may help you plan your revision:

Topic	Revised
• Ecosystems and populations (including practical work)	
• Respiration and photosynthesis	
• Energy transfer and trophic levels, etc.	
• Carbon cycle and nitrogen cycle	
• Man's effect on the planet	
• Development of ecosystems	
• Evolution of new species, isolation, gene pools, etc.	
• Mendel's genetics, populations and Hardy-Weinberg	

About this Unit

Most of this unit is about organisms and their surroundings – a topic often called *Ecology*. It can be thought of as 'whole-organism' biology rather than just a study of what goes on inside organisms. There's also a section on genetics, building on what you learnt about DNA and variation at AS level.

What the examiners are looking for

It's important to know what skills the examiners must test. You can find them in an often-overlooked section of the specification called *Assessment Objectives* (AOs).

To save you looking, they are:

- AO1. 30% of marks. **Knowledge and understanding of science and How Science Works**. Questions covering this objective will ask you to recognise, recall and show understanding.
- AO2. 40% of marks. **Application of knowledge and understanding of science and How Science Works**. Here you'll find words such as analyse, evaluate, apply and assess. These are more advanced skills than simply 'learning the stuff' and are not something that you can put off until revision time. You need to practise them throughout the course so that they become second nature.
- AO3. 30% of marks. **How Science Works**. This is largely the practical work.

So, in the A2 course more weighting is given to AO2 than to AO1. This means that it's very important from the start to practise the interpretation and analysis of data, including the use of statistical tests. Many of the questions at the end of the book give you an opportunity to practise these skills.

3.4.1 The dynamic equilibrium of populations is affected by a number of factors.

Studying populations and ecosystems

You will need to know the following important definitions.

Definitions

Populations and ecosystems: some key definitions

Ecosystem – *a natural unit consisting of producers, consumers and decomposers together with non-living components, for example, a pond, lake, coral reef or rainforest. The conditions within a particular ecosystem are usually fairly uniform.*

Population – *a group of individuals of the same species. The range of the population varies according to the species; the water fleas in a pond constitute a population, but so does the entire human population on Earth. Importantly, members of the same population can potentially interbreed.*

Community – *all of the organisms of all species in the ecosystem. The communities found in a particular habitat are based on dynamic feeding relationships. This means that the size of a population is determined by other populations that it preys on, or that prey on it.*

Habitat – *an organism's environment. For small organisms the immediate surroundings – the microhabitat – is often of vital importance. For instance, aphids (for example, greenfly) can usually be found on the underside of a leaf, next to a vein. If the individual moved just a millimetre away the conditions would change – there could be less food available, and the greenfly may be more exposed to wind movements.*

Niche – *a concept that explains an organism's place in the ecosystem. A niche is largely defined by what an organism eats (unless it is a plant), what eats it and what conditions it lives in. The* **competitive exclusion principle** *states that no two species can occupy precisely the same niche, so they don't compete for precisely the same resources.*

Ecosystems are unbelievably complex, and to even begin to understand what is going on we must take careful measurements, both of the organisms in an ecosystem and of the physical/chemical conditions that form their habitat. In this section, we look at some methods for sampling organisms and some techniques used to measure **abiotic factors** (physical, or non-living, features of the ecosystem).

Abiotic factor	Measured by	Effect of abiotic factor on organisms
Temperature	Thermometer/ thermal probe	Enzymes and therefore metabolism are temperature sensitive. Metabolic systems only work efficiently within relatively narrow temperature ranges. If the temperature is too low, metabolism slows; if it is too high, metabolism becomes imbalanced, and in extreme cases enzymes can be denatured.
Humidity	Hygrometer (a hand-held device)	Affects the rate of evaporation; the higher the humidity, the lower the rate of evaporation. The level of humidity determines the speed of evaporation, and so controls the effectiveness of transpiration in plants and thermoregulation (sweating/panting) in animals.
Light intensity	Light meter or light sensor	Often a limiting factor in photosynthesis – which affects the productivity of the whole ecosystem.
pH	pH meter or chemical (indicator) test	Enzymes are very pH sensitive – metabolism can be disrupted if conditions become too acidic or alkaline.
Oxygen concentration in water	Oxygen-sensitive electrode or chemical test	Oxygen is essential for aerobic respiration. Oxygen solubility in water is low and varies with temperature; the lower the temperature, the more oxygen will dissolve in it.
Carbon dioxide concentration	Gas analysis	Essential for photosynthesis – can be a limiting factor in some circumstances.
Wind speed	Anemometer (hand-held device)	Affects rate of evaporation and cooling, so has an impact on transpiration in plants and thermoregulation in animals.

Table 1
Methods of measuring some abiotic factors

Sampling techniques

There are two basic techniques commonly used in field studies: **quadrats** and **transects**. Quadrats allow you to sample different areas, while transects allow you to measure change from one area to another.

Quadrats

Frame quadrats are sample areas of ground that are small enough to be studied in a short time. A square frame of 50 cm × 50 cm is often used because the grid is manageable and portable (Fig 1a). Quadrats are normally used to compare one area of ground with another, such as the vegetation on north- and south-facing sides of a hill, or the species diversity on mown and unmown patches of ground. You obviously cannot count all the plants in the area, but by placing quadrats randomly you can sample a representative area. Methods of ensuring that the quadrats are random – i.e. without human bias – include mapping out the area into a grid pattern and selecting squares using random numbers from tables or from a computer program designed for this purpose.

Transects

Line transects and belt transects (see Fig 1b) are used to sample organisms along a line in order to show a change from one area to another, such as down a rocky shoreline or along sand dunes.

Examiners' Notes

In examination answers, throwing quadrats is not a satisfactory way of placing quadrats randomly (even if that's what you did on your field trip). You must say that you mapped out the area, created a grid and selected squares without bias by using a table of random numbers or a specific computer program.

a

Fig 1

a Quadrats are usually used to sample different areas of ground; the area can be divided using a grid and quadrats are then placed in squares chosen at random. The experimenter may measure the occurrence (*Is the species present or not?*), number of individuals or the percentage cover of the different species. Percentage cover is most easily done using a quadrat that is divided into 100 smaller squares (perhaps 5 cm × 5 cm) and counting the number of squares in which the species is present

b

Point transect: designed to record vegetation at a series of points

quadrat frame

plants

Belt transect:
measures quantities of vegetation (within quadrat frames)
(a) shows an interrupted belt transect
(b) shows a continuous belt transect
(c) shows a point transect

b Transects are lines that allow us to sample along a changing habitat. Different types of transects include the **interrupted belt transect**, where quadrats are placed at intervals along the line, a **continuous belt transect**, which is self-explanatory, or a **point transect** where just the species touching a particular point on the line are recorded.
NB: A point transect and a line transect are both transects that are done without the use of quadrats. However, with a line transect you can record every species that touches the line, while with a point transect you record the species that touches specific points (say every 25 or 50 cm)

The most appropriate size of quadrat depends on the nature of the area being studied. A rainforest may require quadrats of 20 m square (achieved by pegging out rope/string), while lichens or mosses on a stone wall or tree trunk might only need a quadrat of 25 cm square.

Mark-release-recapture

Transects and quadrats are not a lot of use for estimating the population of animals that can move. It might work for limpets, but trying to throw a quadrat on a rabbit is a fruitless task. We need to be a little more devious.

1 Capture a number of individuals using a suitable trap (from those shown in Fig 2). Longworth traps are good for small mammals such as mice and voles. Pitfall traps may be more suitable for ground-dwelling insects.

Examiners' Notes

Make sure you know the difference between **qualitative data** (in this context: what species are present) and **quantitative data** (how many individuals are present, or the percentage cover/relative abundance).

Remember that when collecting data, random sampling is important because it results in data which are unbiased and therefore suitable for statistical analysis.

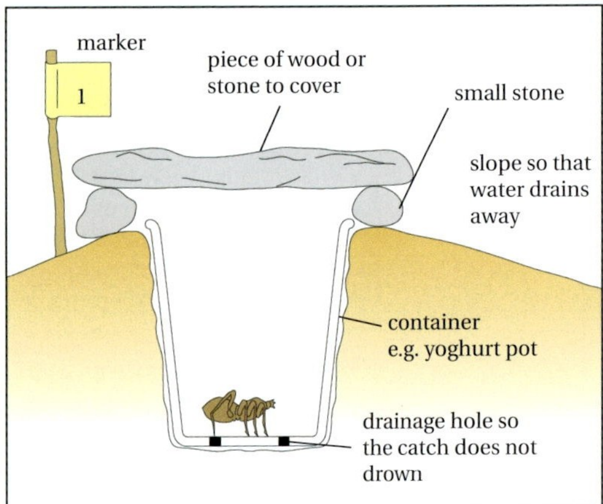

Pitfall traps can be used to trap invertebrates that are active on the soil surface or in leaf litter. 10% methanal (formalin) can be placed in the pitfall to kill predators that might otherwise kill other captives. Pitfalls are cheap and easy to use, but the number of individuals caught tends to reflect the activity of a particular species as well as its abundance.

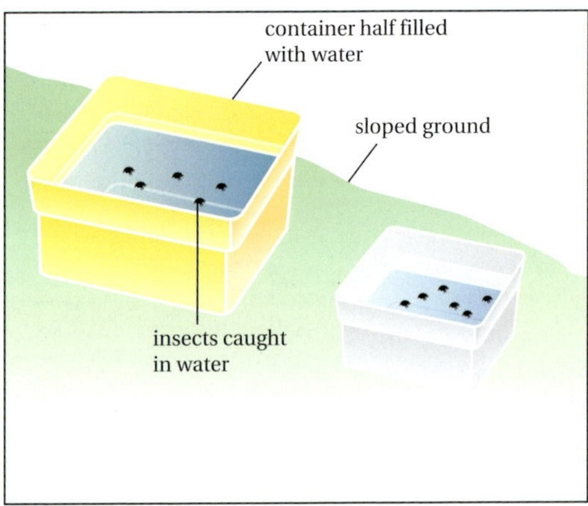

Water traps are left on open ground at different heights. Yellow traps seem to attract aphids while white attracts flies. They can be made from old ice cream cartons, half filled with water. Some washing up liquid can be added to reduce the surface tension so that insects landing on the water will sink.

Source: Adapted from Wiltshire Wildlife Trust

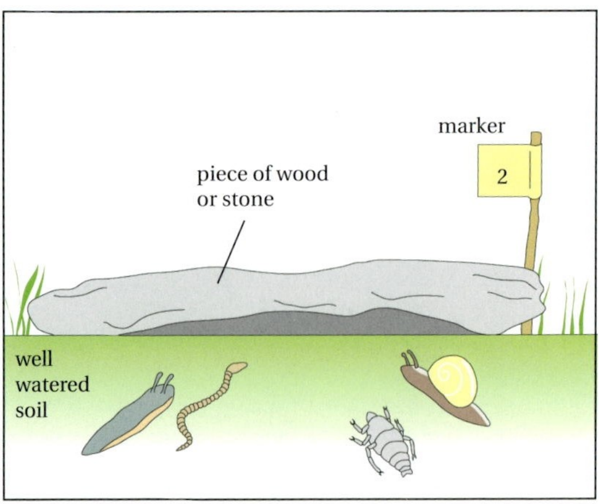

Cover traps are left for a few days before inspection. Like pitfall traps they can be baited with meat, jam or potato. The catch includes slow-moving animals like slugs, earthworms, snails and woodlice.

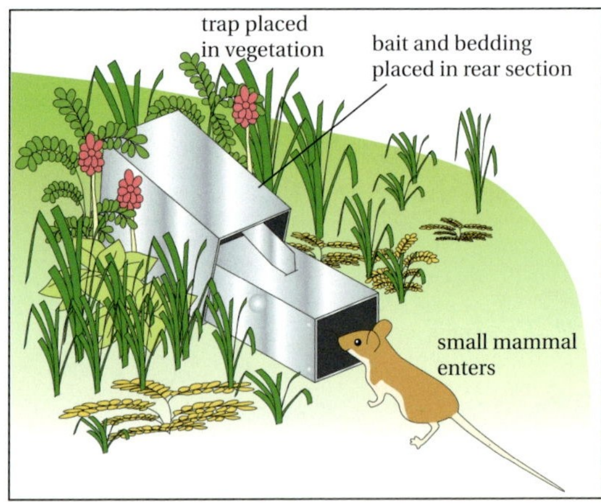

Longworth traps are baited with cheese or other food, and left in long grass where small mammals are likely to be found. As the animal enters, it triggers a lever which allows the trap door to close behind it. The trap should contain warm, dry bedding so that the animal comes to no harm before the trap is revisited and the animal is released.

Fig 2

Different types of traps. Ethical issues here include the fact that water traps kill most of the organisms that fall into them, and that shrews in Longworth traps may starve to death in a very short time

2 Mark the individuals so that you will recognise them if you catch them again. Obviously, the animals should not be harmed, nor marked in such a way that they are less mobile or more visible to predators. (Marking snails with fluorescent dye may seem like a good idea at the time, but ...)

3 Release the marked animals.

4 Leave enough time for the marked individuals to redistribute themselves among the unmarked population (see formula at the top of the next page), then capture a similar number and count the number of individuals that are marked and have therefore been caught before.

The population size can be estimated by the following formula, known as the **Lincoln index**:

$$\text{Population} = \frac{M \times C}{R}$$

Where:

M = total number of animals captured and marked on the first visit

C = total number of animals captured on the second visit

R = number of animals captured on the first visit that were then recaptured on the second visit (i.e. number in second sample that were marked).

Worked Example

On one night 90 bank voles, *Clethrionomys glareolus*, were captured and marked by cutting their fur in a small area (so the dark fur underneath showed up). On the second night 85 were captured, of which 14 were marked.

$$\text{Population} = \frac{90 \times 85}{14} = 546 \text{ individuals (rounded to nearest whole number)}$$

To put this into perspective, consider the two extremes. If you capture 100 on the first night, and 100 on the second night, of which just one is marked, the population is about 10 000. (If none is marked, the formula will not work and the population is too large to estimate.) However, if you capture 100 on the second night that are all marked, then you have probably recaptured the entire population, which must be 100. Of course, this is unlikely.

Essential Notes

The **mark-release-recapture** method assumes that the population is stable, and that no births/deaths or emigration/immigration took place between the first and second samples. The calculation shown in the worked example is probably the simplest one that can be done. Repeated measurement over time, especially when animals are tagged/ringed and given individual numbers, can give valuable information about population size, long- and short-term change, migration patterns, etc.

Risk management

Any field work must be conducted safely, which usually involves some basic precautions. A risk assessment is needed before field work is undertaken.

Higher risk activities include:

- working near water, in remote countryside, wintery conditions, on or near cliffs or steep terrain
- working in an area where extremes of weather or sudden environmental change can occur
- overseas field work, which of course presents a greater risk in terms of more dangerous wildlife, exposure to more potentially fatal diseases and the availability of health care.

Water presents particular problems. Areas of risk include:

- slippery and sharp rocky shore work, compounded if the sea is rough because students may get swept off the shore by an abnormally large wave
- fast-flowing streams and rivers where students can be swept off their feet
- deep water where students could drown
- ponds or streams with steep and/or slippery banks
- polluted water – even a relatively clean looking stream or pond can contain a variety of dangerous bacteria.

Essential Notes

Weil's disease is caused by bacteria transmitted through rat's urine. Water polluted by farmyard waste, sewage and similar forms of effluent may be infected. Weil's disease can be very dangerous but the risk of contracting it from doing field work in fresh water is negligible provided that appropriate precautions are taken.

Reasonable precautions

- Only work in places where the risks of falling in and of pollution are small and where it is shallow enough so that anyone falling in is unlikely to swallow water.
- Ensure that cuts or other forms of broken skin are covered by a plaster or gloves before the field work begins.
- Ensure that all participants - students and adults - wash their hands immediately afterwards. Packed lunches should be consumed before starting field work in fresh water.

Ethical issues

The major ethical issue with field work is that the ecosystem being studied should not be damaged. Large numbers of people all studying the same habitat can leave the populations of some species permanently damaged. For example, a whole year group of students performing a kick sample in a river or stream can seriously disrupt the population of many invertebrates. This in turn disrupts the whole food chain.

Overall:

- Students are expected to be aware of how they can cause minimum impact to the habitat and the organisms they are studying.
- There are opportunities to learn about and debate the conflict between human interest and the environment, including the pros and cons of habitat management and the importance of environmental sustainability.

Populations

Variations in population size

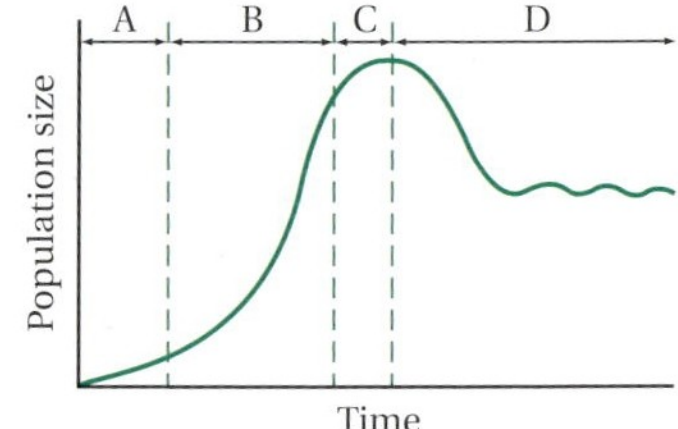

Fig 3a
A generalised population growth curve

Suppose you introduce a pair of rabbits onto an island. How would the population grow? Assuming that they were a healthy pair, and they managed to reproduce, the population growth would be like that shown in Fig 3a. This classic pattern is seen in many different situations; rats in a warehouse, beetles in a sack of flour, elephants in a game reserve, bacteria in your navel or humans on the planet; only the timescale changes. The size of a population depends on a combination of **biotic** and abiotic environmental factors.

Definitions

Size of a population: some key definitions

Abiotic factors *– the non-living factors that affect an organism. These include light intensity, temperature, wind movement, pH, humidity and many others.*

Biotic factors *– the living factors that affect an organism; for example, food supply, predation, competition (from other species and from individuals of the same species) and disease.*

Density-dependent factors *– factors that change with population size. Food supply, for example, is usually density dependent – the larger the population, the greater the competition for food.*

The stages in the population growth curve (Fig 3a) are:

A **Lag phase** – a time of slow growth. There are many different reasons for the lag; microorganisms may need time to activate genes and synthesise the enzymes needed to utilise a new food source, and in more complex organisms, species that reproduce sexually may take a while to grow and reach sexual maturity.

B **Log (logarithmic)** or **exponential phase** – a period of rapid and unrestricted growth, when conditions are favourable, for example, when there is plenty of food. The key point is that there are no **limiting factors**.

C Growth slows due to limiting factors. No population can go on increasing indefinitely; sooner or later there will be **environmental resistance** of some sort. Food may become scarce, waste may accumulate, etc.

D The population stabilises at its **carrying capacity**; the size of population that can be supported in a given area, for example, a pond may be able to support a population of 2 000 water fleas, but not more.

Examiners' Notes

In questions about limiting factors, think carefully about the species involved. Organisms seldom run out of space. Food and water supplies tend to run out long before organisms are standing shoulder to shoulder.

Factors affecting populations

No matter what the species, no population can go on expanding indefinitely. The human population has been rising for thousands of years but even we cannot go on in the same way. If we carried on reproducing to our fullest capacity, a point would be rapidly reached when there was not enough food, too much overcrowding and disease. At the moment there are different rates of population growth in different parts of the world for various social and economic reasons (see page 14).

Most populations are limited by a simpler set of factors than those that affect humans:

- **Competition** – all organisms are locked into a struggle to eat and not be eaten until they have a chance to reproduce. More organisms are born than can possibly survive and an inevitable consequence of this is competition. Common examples of competition include plants competing for light, water and minerals from the soil, while animals may compete for food, mates or nesting sites. There are two types of competition:

  – **Interspecific** competition occurs in individuals from *different* species. Badgers and foxes, for example, may compete for some of the same food sources and burrows. Some plants secrete chemicals that inhibit the growth of competing species.

  – **Intraspecific** competition occurs in individuals of the *same* species.

- **Predation** – this is also a key biotic factor. The populations of predator and prey are often closely linked (Fig 3b), especially if the predator preys on one particular species, as in the case of ladybirds and aphids.

Examiners' Notes

To help you remember which type of competition is which: *inter*specific competition is *between* species like *inter*national football matches are *between* nations.

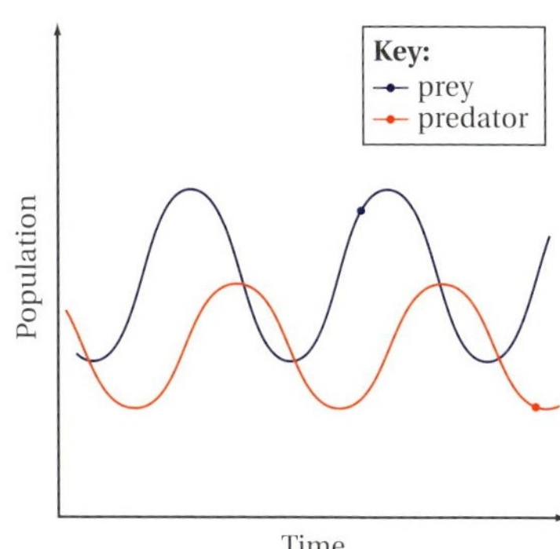

Fig 3b
A simplified graph to show the relationship between predator and prey. When the prey population is high, there is a lot of food for the predators, whose population rises after a time lag. The large predator population takes more of the prey animals, whose population falls; consequently, there is less food for the predators, whose population then falls

Essential Notes

Predator–prey relationships such as the one in Fig 3b are only this clear when the predator relies on one particular prey. Most relationships are more complex than this, as predators have more than one prey animal.

Human populations

Definitions

Human populations: some key equations

***Population growth** can be calculated from:*

- *number of births plus immigration*
- *number of deaths plus emigration.*

So:

population growth = (births + immigration) – (deaths + emigration)

$$\text{percentage growth rate} = \frac{\text{population change during period} \times 100}{\text{population at start of period}}$$

$$\text{birth rate} = \frac{\text{number of births per year} \times 1\,000}{\text{total population that year}}$$

$$\text{death rate} = \frac{\text{number of deaths per year} \times 1\,000}{\text{total population that year}}$$

Essential Notes

A multiplier of 1 000 is used because birth and death rates are reported per thousand of the population.

The age structure of a country can be shown in a **population pyramid** (Figs 4a, 4b and 6), which usually splits the population into five-year segments and shows the percentage of males and females in each age group. The shape of the pyramid tells us a lot about the wealth and health of the nation, and the population's stage of socio-economic development.

A **survival curve** (see Fig 5) shows the number of survivors from a sample of (usually 10 000) people plotted against time. **Average life expectancy** is the age at which 50% of the population in the sample used are still alive.

Fig 4a
Population pyramids for a less economically developed nation (LEDN) and a more economically developed nation (MEDN)

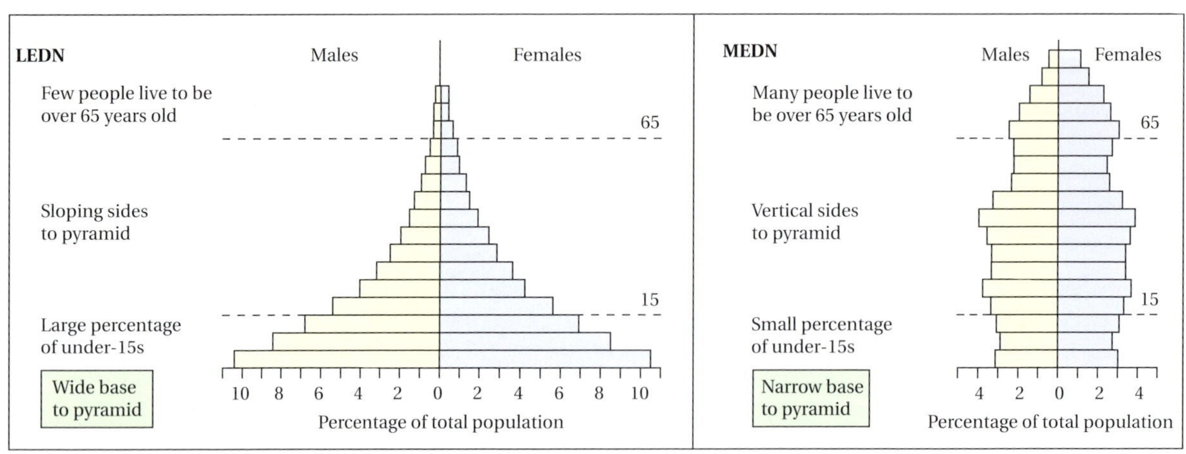

Source: Adapted from Hornby and Jones, *Introduction to Population Geography*, Cambridge University Press, 1993

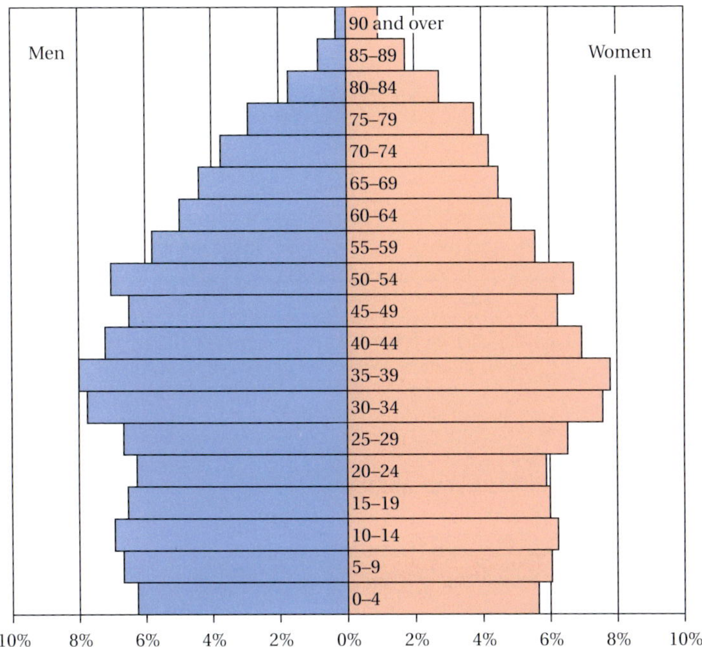

Fig 4b
Age pyramid for the UK in 2006; the UK has a population of just over 60 000 000, with about 500 000 deaths per year and about 650 000 births

Demographic transition and survival curves

Demographic transition means a change in the population structure.
Fig 5 shows a variety of human survival curves for the UK, showing that more and more people are surviving into old age. You need to be able to interpret these curves and suggest reasons for the changing patterns.

Demographic transition is a model used to explain the social and economic development of a country from a pre-industrial to an industrialised economy. The central idea is a shift from high birth rates and high death rates to low birth rates and low death rates as a result of improved education, sanitation (access to clean water), health care and diet.

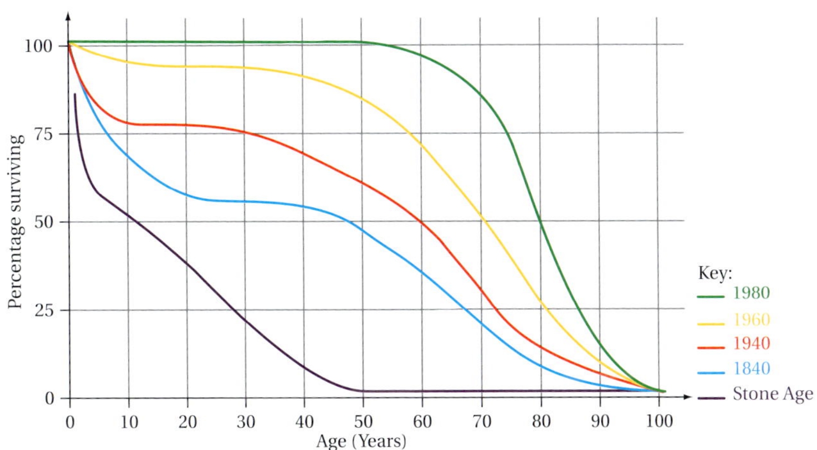

Fig 5
Survival curves for the UK through the ages. The obvious difference is that more and more people are surviving into old age

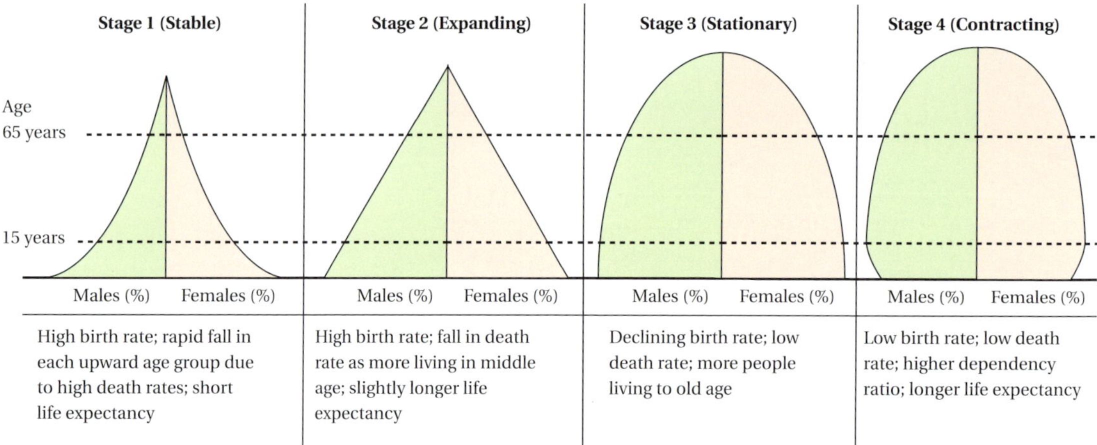

Fig 6
The four stages of demographic transition

Stage 1 (Stable) – high birth rate and high death rate, typical of a pre-industrial nation. Seen in the UK 200–300 years ago, and still seen today in some of the poorest countries. Note the high infant mortality and comparatively few individuals surviving to middle/old age.

Stage 2 (Expanding) – high birth rate but a lower death rate which leads to an expanding population. Better sanitation, health care and diet means that more individuals survive into middle age and beyond.

Stage 3 (Stationary) – lower birth rate and lower death rate, so the population begins to stabilise. Birth rate falls due to access to contraception, increases in wages, an increase in the status and education of women, children staying longer in education rather than leaving early to find work.

Stage 4 (Contracting) – birth rate drops to below death rate, leading to a shrinking population, as currently seen in Germany, Japan and Italy.

3.4.2 ATP provides the immediate source of energy for biological processes.

The relationship between photosynthesis and respiration

From GCSE, you should remember the basic equations for these processes.

Photosynthesis:

$$\text{carbon dioxide + water} \xrightarrow[\text{chlorophyll}]{\text{light}} \text{glucose + oxygen}$$

Respiration:

$$\text{glucose + oxygen} \rightarrow \text{carbon dioxide + water + energy}$$

These two processes are effectively the reverse of each other. Photosynthesis traps energy in organic molecules and this energy is released in respiration. Overall, photosynthesis is the *reduction* of carbon dioxide to form organic molecules, while respiration is the *oxidation* of organic molecules to form carbon dioxide.

What is reduction?

In reduction reactions, substances gain electrons. Electrons contain energy, so a substance that is reduced gains energy. When carbon dioxide is reduced to form carbohydrates such as glucose, the carbon dioxide molecule *gains* energy.

What is oxidation?

The opposite of reduction; oxidation is a reaction in which substances lose electrons and therefore lose energy. In respiration, organic molecules such as glucose have the electrons stripped from them, and the energy released is used to make ATP.

Facts about ATP

- **ATP** stands for adenosine triphosphate. It can be **hydrolysed** (broken down) into adenosine diphosphate (ADP) and free inorganic phosphate (P_i) (Fig 7).
- Splitting ATP provides readily available energy in small, usable amounts for the wide variety of energy-requiring reactions that occur in cells.
- ATP is a relatively small molecule that can diffuse around the cell quickly.
- ATP is an intermediate energy source within cells. It carries energy from the site of respiration, mainly the **mitochondria**, to other areas of the cell where energy is needed.

In respiration, the energy contained in organic compounds is released in a series of steps. This energy is used to make ATP from ADP and phosphate. Respiration can therefore be thought of as a process that replenishes ATP stocks, making sure that it is made as fast as it is used.

The ATP, ADP and P_i in a cell are sometimes known as the 'phosphate battery'. If all of the phosphate were to exist as ATP, the battery would be 'fully charged'. However, ATP is an unstable molecule, and is constantly being broken down and re-synthesised in a cycle within the cell.

Examiners' Notes

Be careful how you use the term 'energy'. The process of respiration *releases* the energy in organic molecules, and *transfers* it to ATP. Candidates often lose marks when they state that respiration *makes* or *produces* energy. Energy cannot be made or destroyed.

Examiners' Notes

ATP and energy are not the same. You cannot say that ATP is energy – ATP is a molecule that releases energy when it is hydrolysed to form ADP and phosphate.

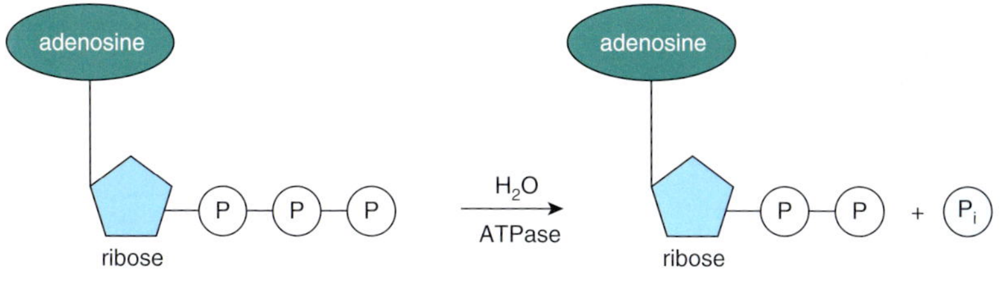

Fig 7
The hydrolysis (splitting) of ATP. This reaction releases energy that can be used to drive energy-requiring reactions, such as muscular contraction

3.4.3 In photosynthesis, energy is transferred to ATP in the light-dependent reaction and the ATP is utilised in the light-independent reaction.

Photosynthesis

Photosynthesis is often seen by students as a difficult topic to learn, but you can make sense of it all if you approach it in the right way. First, you need an overview; this provides a framework on which you can fit the details. The best way to start is to divide photosynthesis into two easy steps.

1 **The light-dependent reaction** – light hits **chlorophyll**, which then emits high-energy electrons. These electrons pass through a series of electron transfer reactions that make ATP and NADPH. The electrons in chlorophyll are replaced when water is split, a process that also produces oxygen as a by-product.

2 **The light-independent reaction** – ATP and NADPH are used in a series of reactions known as the **Calvin cycle** (Fig 9). Overall, these reactions reduce carbon dioxide to glucose.

Both steps take place in the **chloroplast** (Fig 8), but it is important to remember that different reactions happen in different parts of this important organelle.

The light-dependent reaction can be outlined as follows:

- This process happens on the **thylakoids**, which are membranes within the chloroplast that contain the chlorophyll molecules.
- When a photon of light hits a molecule of chlorophyll, the pigment becomes *excited* (raised to a higher energy level) and emits two high-energy electrons.
- These excited electrons pass down an **electron transport system** on the thylakoid membrane. The result is ATP synthesis by **photophosphorylation**.
- Energy is also used to split water via the process of **photolysis**, which produces new electrons to replace those lost by the chlorophyll. Protons (hydrogen ions) and oxygen are made as by-products.
- Electrons are used to make reduced NADPH (also called NADP).
- ATP and NADPH are essential for the light-independent reaction, and oxygen is released into the atmosphere.

Essential Notes

NADPH is similar to NADH (seen in respiration) and performs basically the same electron-carrying function. It is useful to remember 'P for photosynthesis'; NADH in respiration, NADPH in photosynthesis.
(**NB**: in reality, the 'P' stands for phosphate.)

Examiners' Notes

A common mistake is to state that starch in a chloroplast provides extra energy for photosynthesis.

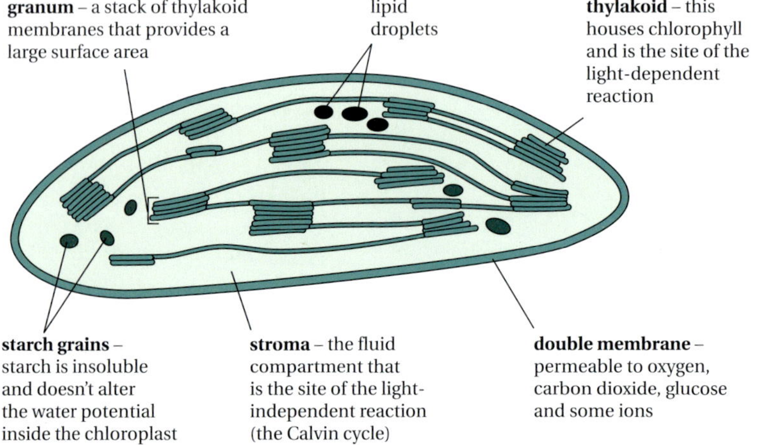

Fig 8
Chloroplasts are organelles that house all the enzymes and other substances needed for photosynthesis; they take the form of biconvex discs (rather like Smarties™ – but don't call them that in the examination!)

The light-independent reaction can be outlined as follows:

- This process happens in the **stroma**, the fluid in the centre of the chloroplast.
- It involves the reduction of carbon dioxide by a series of reactions known as the Calvin cycle. The essential stages of this cycle are shown in Fig 9.
- ATP and NADPH are used to reduce the carbon dioxide. Supplies of these compounds depend on light, so, when it gets dark, the light-independent reaction finishes soon after the light-dependent reaction.

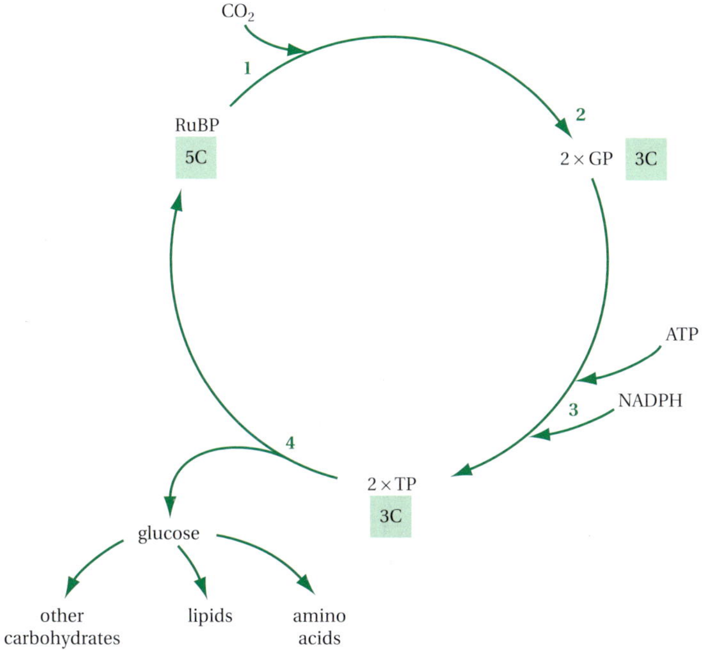

Fig 9
The Calvin cycle

Essential Notes

TP (triose phosphate) is sometimes referred to as GALP (glyceraldehyde phosphate).

The four key stages of the Calvin cycle can be described as follows:

1 Carbon dioxide combines with the 5-carbon compound **ribulose bisphosphate** (**RuBP**). This reaction is catalysed by the enzyme **rubisco** (ribulose bisphosphate carboxylase).

2 This produces a highly unstable 6-carbon compound that immediately splits into two molecules of **glycerate 3-phosphate** (**GP**, a 3-carbon compound).

3 ATP and NADPH are used to reduce the GP into **triose phosphate** (**TP**, another 3-carbon compound), sometimes referred to as GALP (glyceraldehyde phosphate). This is the first compound that is actually a sugar.

4 Some of the triose phosphate is used to make carbohydrate, but most of it is used to make more RuBP to continue the cycle. For every glucose molecule produced, five molecules of RuBP are re-synthesised. Glucose can be converted into many other essential organic compounds. As well as the comparatively simple conversion into starch, glucose can also be converted into triglycerides and, using phosphate and nitrate ions from the soil, phospholipids, proteins and nucleic acids.

Limiting factors in photosynthesis

The limiting factor is the one in shortest supply. Increase the supply and you will increase the rate of reaction. Common limiting factors of photosynthesis include temperature, carbon dioxide concentration and light intensity. For example, at night, light is obviously a limiting factor. At dawn, the light intensity increases and so does the rate of photosynthesis until some other factor – possibly carbon dioxide level – becomes limiting.

Temperature is often a limiting factor. The light-dependent reaction is not particularly temperature sensitive because it doesn't rely on enzymes. Instead, it relies on the excitation of chlorophyll followed by electron transport chains. In contrast, the light-independent reaction is much more temperature dependent because it is controlled by enzymes.

A knowledge of limiting factors and how to increase their supply is obviously important in agriculture (Table 2).

Table 2
Improving the abiotic environment of crop plants

Factor	How supply can be increased	Is it worth it ...	
		in a glasshouse?	in a field?
Temperature	Heater	Possibly	No
Carbon dioxide	Heater – burn a fossil fuel (for example, propane)	Possibly	No
Mineral ions	Add fertiliser	Usually	Usually
Light	Artificial lighting	Possibly	Rarely
Water	Spray or irrigate	Essential	Yes, if rainfall/ irrigation is insufficient

Measuring photosynthesis

A common way to measure the rate of photosynthesis is to use an aquatic plant such as *Elodea*, and measure the rate of oxygen consumption. The plant will produce bubbles of oxygen which can be collected (Fig 10).

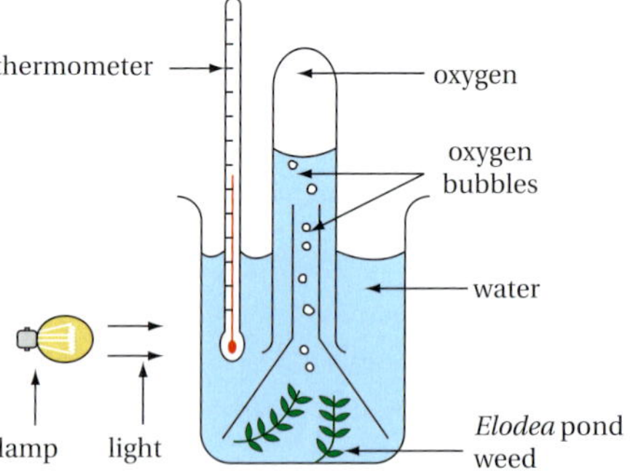

Fig 10
Apparatus used to measure the rate of photosynthesis in *Elodea* pond weed

This apparatus can be used to investigate the effect of various environmental factors. Light intensity, temperature and hydrogencarbonate (dissolved CO_2) concentration can all be used as independent variables. A common problem is the need to perform repeat experiments, given that no two pieces of pond weed will be identical. To allow differently-sized specimens to be compared, the surface area of each plant used can be estimated (which is messy, and usually involves each leaf being removed and its area estimated using graph paper). This allows the rate of photosynthesis to be given as 'volume of oxygen per unit time per unit surface area of leaf'. For example, 5 cm^3 O_2 min^{-1} cm^{-2}.

Are we really measuring the rate of photosynthesis?

The simple answer is 'no', because we haven't taken **respiration** into account. The amount of oxygen a plant gives off is a measure of how much the rate of photosynthesis *exceeds* the rate of respiration. The actual rate of photosynthesis can be estimated if the experiment is repeated in the dark.

For example, a student measured the rate of photosynthesis in the light and the rate of respiration in the dark (Table 3).

Process	Volume of oxygen used or made (cm^3) per minute
Respiration in dark	3 used
Rate of photosynthesis in light	14 made
True rate of photosynthesis	**17**

Table 3
Results of experiment to measure the rate of photosynthesis in the light and rate of respiration in the dark

3.4.4 In respiration, glycolysis takes place in the cytoplasm and the remaining steps in the mitochondria. ATP synthesis is associated with the electron transfer chain in the membranes of mitochondria.

Respiration

Respiration is the release of energy from organic molecules. In this section, we study the breakdown of glucose. This is the main fuel for respiration in humans, but many other organic compounds can also be respired. For example, we respire lipids when our carbohydrate stores run low, and many carnivorous animals, such as cats, get much of their energy by respiring amino acids from their high-protein diet.

In practice, full **aerobic respiration** consists of four processes. These are shown in Fig 11 overleaf, and are described below.

1 **Glycolysis** – one molecule of glucose is split into two molecules of **pyruvate**.

2 **Link reaction** – pyruvate is converted into **acetate**, which then combines with **co-enzyme A** (or **CoA**) to become **acetyl co-enzyme A** (often shortened to **acetyl CoA**).

3 **Krebs cycle** – electrons are removed from the acetyl CoA.

4 **Electron transport system** – the energy in the electrons is used to make large amounts of ATP.

Respiration is an energy-releasing process that takes place in virtually *all* living cells *all* of the time. In order to understand respiration, you need to know about the co-enzyme, **NAD**, discussed on page 21.

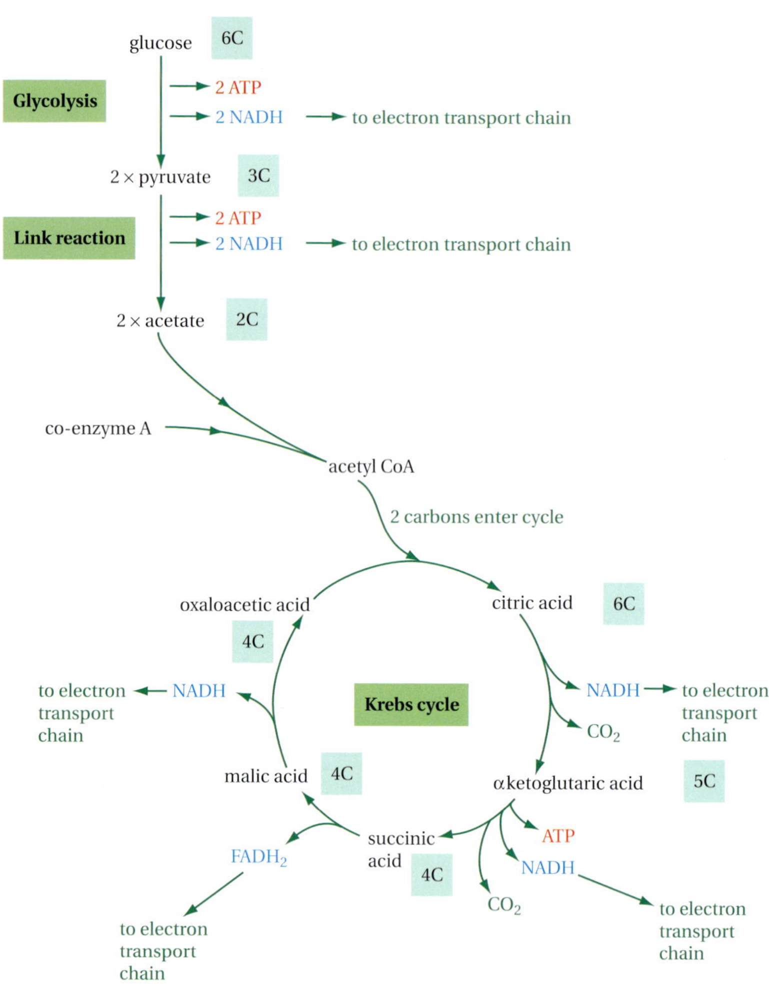

Fig 11
An overview of respiration showing glycolysis, the link reaction, the Krebs cycle and the electron transport chain. The number of carbon atoms is shown in the shaded squares

Electron transport chain (a series of proteins on the inner mitochondrial membrane).

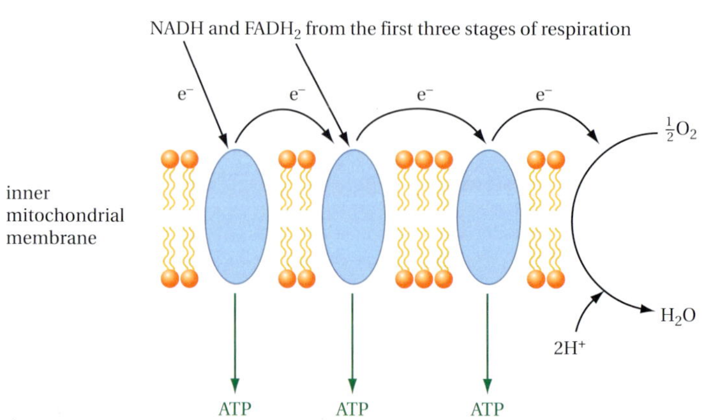

What is NAD?

NAD stands for nicotinamide adenine dinucleotide; although you will not be asked to remember the full name. NAD is a co-enzyme and its key feature is that it carries electrons.

$$NAD^+ + e^- \rightarrow NADH$$

co-enzyme + electron → reduced co-enzyme

Respiration is basically a series of oxidation reactions that remove electrons from the original organic molecule. These electrons are picked up by NAD, forming NADH, also called **reduced NAD**. So whenever you see NADH, think: 'an electron is being carried.'

NADH carries electrons into the electron transport system, where they are used to make large amounts of ATP. $FADH_2$, made in the Krebs cycle, is similar to NADH and has exactly the same electron-carrying function.

The location of each respiration reaction in a mitochondrion is shown in Fig 12.

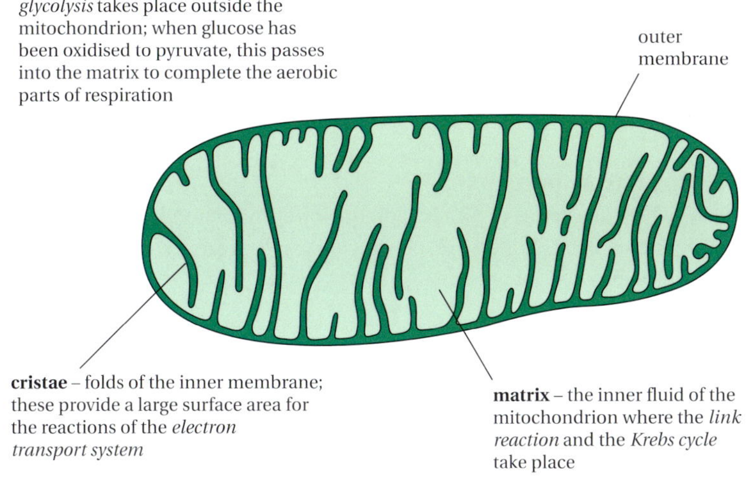

Fig 12
What-happens-where in a mitochondrion

Each of the four reactions of respiration (shown in italics in Fig 12) are looked at in detail below.

Glycolysis

- The word glycolysis means 'sugar splitting.'
- In glycolysis, one molecule of glucose (a 6-carbon compound) is oxidised to produce two molecules of pyruvate (a 3-carbon compound).
- The reaction yields two molecules of ATP and two molecules of NADH for each glucose molecule that is split. (It actually uses two ATP molecules but produces four – a net profit of two.)
- Glycolysis takes place in the **cytosol** – the fluid part of the cytoplasm (*not* in the mitochondria).
- The process does not require oxygen. **Anaerobic respiration** is basically just glycolysis.

Essential Notes

Note the similarities in the structure of a chloroplast and a mitochondrion (Fig 8 and Fig 12). Both are about the same size (roughly 10 μm across) and both have an internal membrane system giving a large surface area for reactions. Both convert energy: chloroplasts convert light energy into chemical energy (by making glucose, etc.); while mitochondria convert one form of chemical energy (glucose, etc.) into another (ATP).

Essential Notes

Respiration and photosynthesis both consist of a series of steps. Each step is a relatively simple chemical reaction that is controlled by a specific enzyme.

Link reaction

This reaction is so named because it links glycolysis to the Krebs cycle.

- This reaction is also called pyruvate oxidation.
- In the link reaction, pyruvate is used to produce acetate and carbon dioxide. The acetate is picked up by co-enzyme A, forming acetyl co-enzyme A.
- No ATP is produced during the link reaction, but two molecules of NADH are formed.
- The link reaction takes place in the **matrix** (inner fluid) of the mitochondria (see Fig 12).

Krebs cycle

- This cycle is a series of reactions that oxidise what is left of the glucose after it has passed through glycolysis and the link reaction. By now, the glucose has been converted to acetate. In the Krebs cycle electrons are stripped from the acetate, creating large amounts of the electron carriers NADH and $FADH_2$.
- The reactions take place in the matrix of the mitochondria.
- Each turn of the cycle produces one ATP, three NADH, one $FADH_2$ and two CO_2 molecules.
- The cycle turns twice per molecule of glucose. It therefore produces two ATP, six NADH, two $FADH_2$ and four CO_2 molecules per molecule of glucose.

The electron transport chain

- This consists of a series of proteins on the inner mitochondrial membrane, which is folded into **cristae** to provide a large surface area for the process.
- Electrons are delivered to the electron transport system by NADH and $FADH_2$.
- The electrons pass from one protein to the next along the chain.
- Each electron transfer is an oxidation/reduction reaction that releases energy.
- This energy is used to pump protons (H^+ ions) from the matrix, across the inner membrane into the outer mitochondrial space by **active transport**.
- This creates a diffusion gradient of H^+ ions, which diffuse back into the mitochondrial matrix through the centre of ATPase enzymes (see Fig 13 opposite). As they do so, the ATPase enzyme synthesises a molecule of ATP.
- The by-products of this process are low-energy electrons and protons, which combine with oxygen to form water.
- You are breathing at this moment because of the electron transport chain. You need to provide oxygen to mop up the electrons, and also to remove the accumulated carbon dioxide from your lungs.

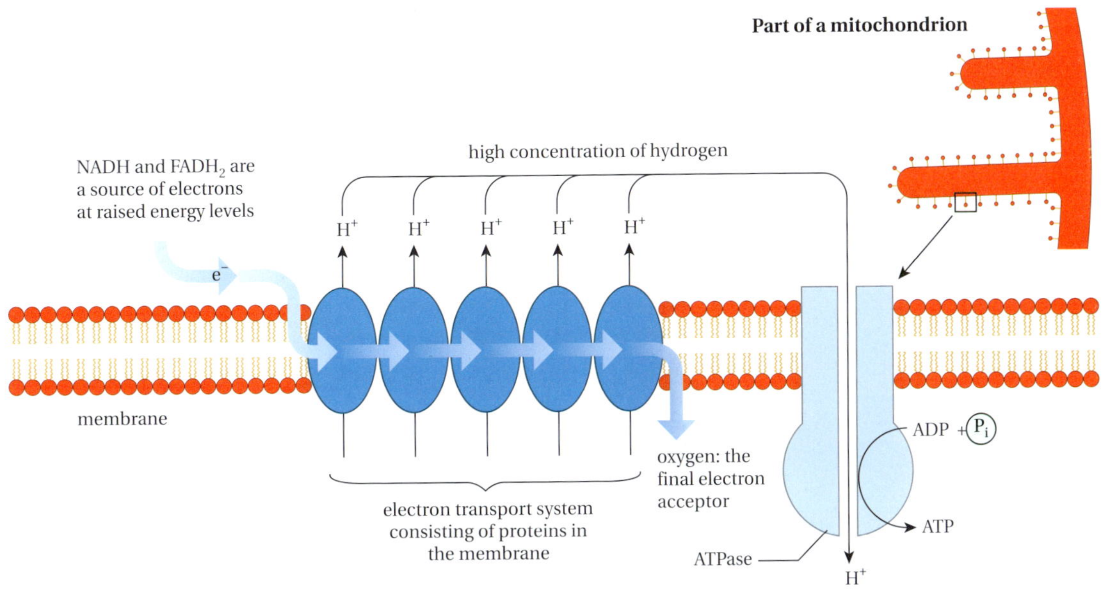

Fig 13
The electrons and hydrogen ions made during the first stages of respiration are finally used to synthesise ATP

How much ATP is made per glucose molecule?

In ideal conditions the theoretical maximum is 38 molecules. Below we see why.

There are two ways of making ATP:

- **Substrate level phosphorylation** – the method of ATP production in glycolysis and the Krebs cycle. Both processes provide two ATP molecules per glucose, giving a total of four ATP molecules by this method.
- By **oxidative phosphorylation** – here, ATP is made using energy released in the electron transfer system. In effect, it is ATP made by 'cashing in' the energy in the electrons carried by NADH and FADH$_2$. The other 34 molecules of ATP made during respiration per molecule of glucose are made by this method (see below).

Process	ATP made	NADH made	FADH$_2$ made
Glycolysis	2	2	0
Link reaction	0	2	0
Krebs cycle	2	6	2
Totals	**4**	**10**	**2**

Table 4
The total ATP and NADH production from the first three parts of respiration, per molecule of glucose

When fed into the electron transport chain, three ATP molecules are made per NADH, and two per FADH$_2$. This gives us:

From NADH: $10 \times 3 = 30$ ATP

From FADH$_2$: $2 \times 2 = 4$ ATP

This gives a total of 34 molecules of ATP made by oxidative phosphorylation. When we add these to the original four produced by glycolysis and the Krebs cycle, we get the 38 ATPs produced per molecule of glucose during aerobic respiration. In practice, slightly fewer ATP molecules are made, for a variety of complex reasons.

Anaerobic respiration

This is respiration without oxygen, and it's a much simpler process than aerobic respiration. It's basically glycolysis that doesn't go any further. Compared to aerobic respiration, anaerobic respiration:

- produces less ATP (two compared to about 38)
- takes place in the cytoplasm, not in the mitochondria
- only takes a short time to complete.

In all organisms, glucose is converted into pyruvate. As a general guide, animals and bacteria convert pyruvate into **lactate**, while plants and fungi convert the pyruvate into carbon dioxide and **ethanol**.

In both cases, the conversion of pyruvate is essential because it re-synthesises NAD^+ from NADH. This is vital because otherwise there would be no more NAD^+ available and so glycolysis could not continue.

Measuring respiration

How do you measure the rate of respiration? You can measure energy production (as heat) or oxygen consumption, but in practice the latter is easier. Fig 14 shows one type of respirometer.

A respirometer works in the following way:

- The organism respires. It takes in oxygen and gives out carbon dioxide.
- Normally, this will not change the volume of the gas in the apparatus, because the carbon dioxide made will replace the oxygen used.
- But, the sodium hydroxide absorbs all of the carbon dioxide, so it's just as if the organism isn't making any carbon dioxide.
- As a result, the volume of the air in the chamber decreases as the organism uses oxygen. This draws the fluid along the tube and the rate of oxygen used per unit time can be measured.
- Suitable units for rate of respiration will include the volume of oxygen used per unit mass of the organism per unit time. Actual units will depend on the timescale and the size of the organism. For example, cubic centimetres of oxygen per gram per hour ($cm^3\,O_2\,g^{-1}h^{-1}$).

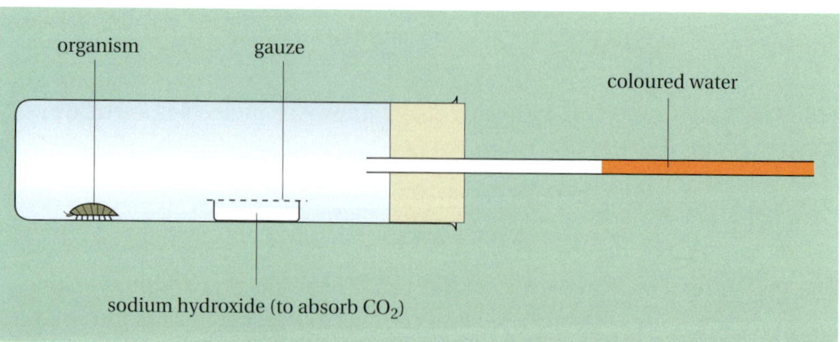

Fig 14
A simple respirometer

3.4.5 Energy is transferred through ecosystems and the efficiency of this transfer can be measured.

The importance of photosynthesis

Photosynthesis is the only process that can capture sunlight energy and so it is the major route by which energy enters an **ecosystem**. In photosynthesis, sunlight is used to reduce carbon dioxide into organic molecules. Initially, simple carbohydrates (sugars) are made, but plants can make other molecules such as lipids, proteins and nucleic acid, by modifying the carbohydrates. In this way, plants make the food molecules that support whole ecosystems. The by-product of photosynthesis is oxygen – another substance vital to life on Earth. Organisms that can photosynthesise are called **producers**. The relationship between producers and other types of organisms in an ecosystem is shown in Table 5.

Table 5
The relationship between producers, decomposers and consumers

Type of organism	What it needs	What it produces
Producer (green plants and algae)	Carbon dioxide, inorganic ions	Oxygen, organic molecules
Consumer (mainly animals)	Organic molecules, oxygen	Carbon dioxide, organic waste
Decomposer (bacteria and fungi)	Organic molecules, oxygen	Carbon dioxide, inorganic ions (nitrate, etc.)*

* Only bacteria produce inorganic ions

Where does the energy from sunlight go?

A huge amount of solar energy reaches our planet but only a small percentage is captured by plants in photosynthesis and packaged into organic molecules. The rest is lost in different ways.

- A large amount of sunlight misses plants altogether – some of it heats the atmosphere; some of it heats the seas and rocks.
- Not all the light that reaches a plant hits the chloroplasts – some passes straight through.
- Some light is of the wrong wavelength – plants use mainly the blue and red light in the visible spectrum, and reflect green. Some energy is used in the evaporation of water from the leaves (transpiration).
- The reactions of photosynthesis, like all reactions, are inefficient – some energy is always lost as heat.
- The term **gross primary production** refers to the total energy in the organic molecules produced by a plant. However, the plant uses a proportion of the energy for its own needs – this energy is released when the plant respires or dies and decomposes. Only the surplus energy produced by the plant – the **net primary production** – is available to the rest of the ecosystem.
- Put as an equation:

$$NPP = GPP - R \quad \text{(where } R = \text{respiration)}$$

Fig 15 shows the energy transfer along a food chain. The Sun may send us a large amount of energy but it is easy to see why it soon runs out. The transfer of energy at each level is very inefficient, usually between 2% and 5%. So the number of steps in a food chain is always limited because there is no energy left. There are usually three steps in a chain; rarely more than five.

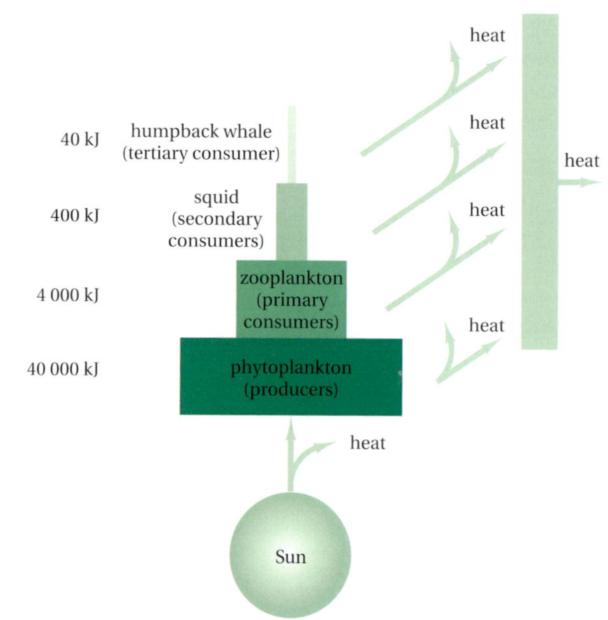

Fig 15
The energy transfer for an Antarctic food chain showing the percentage energy transfer at each level; at each stage, over 90% of the available energy is lost as heat

Fig 16 shows one way to illustrate the different ways in which energy is lost.

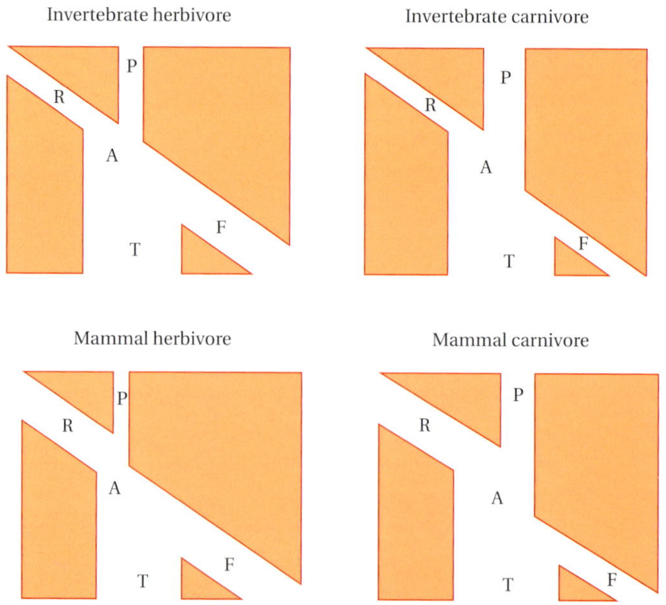

Fig 16
Energy transfer boxes for four animals

T = total energy in the food eaten
NB: food that cannot be digested cannot be absorbed, and so is not available to the organism
A = energy absorbed
F = energy lost in faeces
R = energy lost in respiration
P = energy incorporated into the tissues of the organism; this is the energy passed on to the next **trophic level**

Note two important trends:

1. Mammals pass a smaller proportion of their energy intake up the food chain because they use more energy to maintain their body temperature
2. Carnivores are more efficient at converting their food into body tissue because meat has less indigestible material than plant-based food

Ecological pyramids

Pyramid diagrams show the relative size of different components at each of the trophic levels. Three types of pyramid diagrams are commonly used: *numbers, biomass* and *energy*.

Pyramids of numbers

These refer to the numbers of organisms in the different species in a food chain. Fig 17 shows examples of the main types of pyramids of numbers. Sometimes these are true pyramids – wide at the bottom, becoming narrower towards the top. Other examples have a narrow base, followed by the usual pyramid shape. This happens when the producers are large, for example, trees. A pyramid of numbers can be truly inverted when parasites are involved. For the following food chain, the numbers of organisms may increase at each level:

oak tree → grey squirrel → flea

Pyramids of biomass

Biomass refers to the dry mass of the population within the ecosystem, and so takes into account the different sizes of organisms. There may be millions of aphids on an oak tree, but the mass of the tree far outweighs the mass of the insects feeding on it.

Pyramids of biomass give a more realistic picture of feeding relationships, but they are very difficult or impossible to calculate in practice because biomass refers to dry mass, which you cannot work out without killing the organism. So if you want to work out a pyramid of biomass for an African plain, you have to *estimate* the dry mass of the grass, zebra, antelope, lions, etc. in the food chain.

Pyramids of biomass are usually true pyramids, but there is one classic situation in which the consumers can outweigh the producers: in some areas of the ocean the mass of the zooplankton (minute animals) can outweigh the phytoplankton (minute plants) on which they feed. This is because the producers have a short life cycle and those that escape predation reproduce quickly to replace those that are eaten.

Pyramids of energy

If you calculate the total amount of energy that flows through each trophic level you will always get a true pyramid. Even with the phytoplankton food chain the amount of energy flowing through the producers will be greater than that flowing through the primary consumers.

Examiners' Notes

When you see an inverted pyramid of numbers where there are more secondary consumers than primary consumers, think 'parasites' as an explanation.

Examiners' Notes

Make sure you are clear about how energy is lost. It's basic thermodynamics: energy transfers are never 100% efficient and some is always lost as heat. Movement, for example, involves two energy transfers – from glucose to ATP in respiration and from ATP to kinetic energy in muscular contraction – so there are two opportunities to lose heat.

Examiners' Notes

Make sure you don't make the common examination mistake of stating 'energy is used for respiration'. Some candidates even confuse respiration with photosynthesis.

Food chain	Pyramid of numbers	Pyramid of biomass	Pyramid of energy
fox rabbit grass	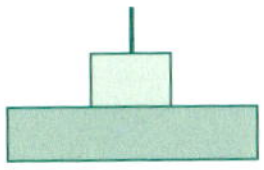	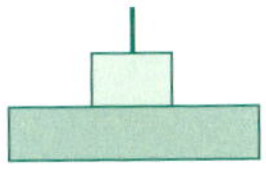	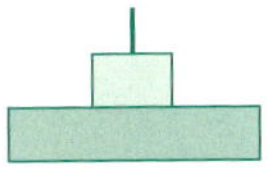
fleas grey squirrels oak tree	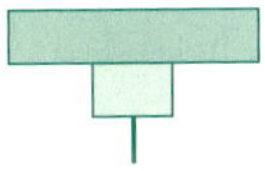	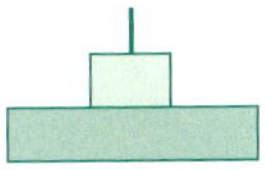	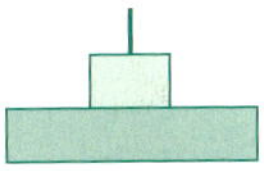
shark herring zooplankton phytoplankton	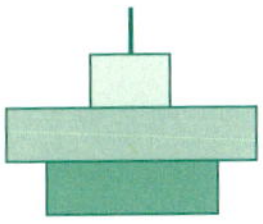	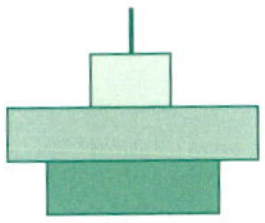	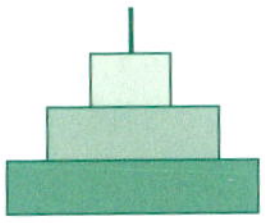

Fig 17
The three types of pyramids for three different food chains

3.4.6 Chemical elements are recycled in ecosystems. Microorganisms play a key role in recycling these elements.

Organisms are made from 'secondhand' material. The atoms and molecules that make up your body have been part of many organisms before you, and will probably be part of many different organisms long after you are gone. These resources are finite, and are recycled again and again. Importantly, energy is not recycled – we need a continued supply in order to drive the cycles, building up simple molecules into more complex molecules and then breaking them down again. You need to know about the **carbon cycle** and the **nitrogen cycle**.

The carbon cycle

Carbon forms the backbone of all organic molecules that make up the bodies of organisms; including carbohydrates, lipids, proteins, nucleic acids and simpler organic molecules.

Overall, the carbon cycle involves carbon dioxide from the atmosphere – or hydrogen carbonate ions (HCO_3^-) in water – being fixed into organic molecules by photosynthesis, and then being released back into the atmosphere by respiration of the various organisms in the ecosystem, as shown in Fig 18.

Fig 18
The two basic steps in the carbon cycle

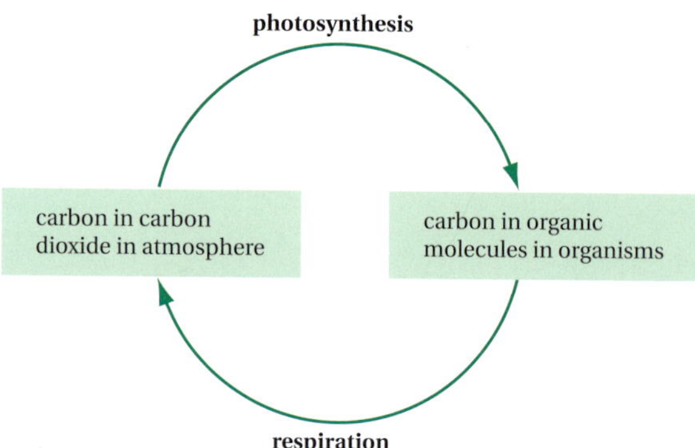

We can build on this simple cycle to show how carbon is passed up the food chain and is released back into the atmosphere by the respiration of the various organisms in the ecosystem (Fig 19). The carbon that is not released by respiration is trapped in fossil fuels such as coal, gas, oil and other deposits (such as peat). The only way this fossilised carbon can be released back into circulation is by **combustion**, which is increasing globally year on year. As we are also reducing the global rate of photosynthesis by deforestation, we are shifting the balance and, as a result, the amount of carbon dioxide – a greenhouse gas – is increasing in the atmosphere.

Global warming

Of the sunlight we receive, some is absorbed by the atmosphere, land and sea; some is reflected out into space. The temperatures on Earth result from a balance in energy received and energy lost. The energy we get from the Sun is not changing, but we are losing less because we are changing the composition of our atmosphere. Water vapour, carbon dioxide, methane and several other gases act together to prevent heat from escaping from our atmosphere.

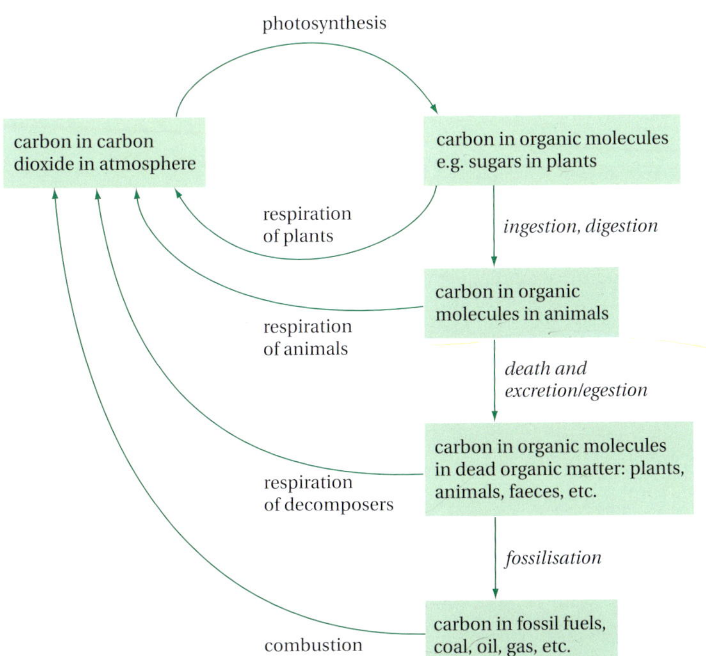

Fig 19
A more detailed carbon cycle. The world's oceans, rivers and lakes contain more carbon dioxide dissolved in water as hydrogencarbonate (HCO_3^-) ions than occurs in gaseous form in the atmosphere. On a global scale, most photosynthesis takes place in the oceans

Why is the atmosphere changing?

There is comparatively little carbon dioxide in the atmosphere; about 0.038% – just 380 molecules in every million (often written as 380 ppm). However, this is at an all-time high and is rising steadily. The main reason for the increase is the burning of fossil fuels. Power stations, industry, car and aircraft engines combine to produce millions of tonnes of carbon dioxide every year.

While there is a massive amount of combustion going on, there is still the process of photosynthesis that absorbs carbon dioxide from the atmosphere. Cutting down forests – **deforestation** – is another matter for concern. There's a lot of carbon locked up in wood, so when we cut down trees and burn them, a lot more carbon dioxide goes into circulation.

Methane gas

Methane gas (CH_4) is another greenhouse gas whose production is increasing. Among the main sources are:

- Fermentation of waste in landfill sites – methane is produced when waste decomposes in anaerobic conditions.
- Natural gas and oil systems – methane losses occur during the production, processing, storage, transmission and distribution of natural gas and oil.
- Flatulent cows – the metabolism of cows and other ruminants (buffalo, goats, sheep and camels) results in large amounts of methane being released.
- Coal mining – a lot of methane is trapped as coal deposits and is released when coal is mined and processed.
- Storage of manure on farms and waste water (sewage) treatment.
- Anaerobic soil, such as rice fields.

The previously-mentioned points are all **anthropogenic** sources, i.e. due to human activity. However, it has been estimated that 76% of the methane in the atmosphere comes naturally from the fermentation of **wetlands** and other areas where conditions are anaerobic.

Global warming is a complex problem

The Earth's climate is a complex system. This means that it's impossible to predict exactly what will happen, but the following are possibilities:

- Higher temperatures make ice melt. Glaciers and other areas of permanent ice will melt and drain into the sea.
- The extra fresh water might disrupt ocean currents. This could be a disaster for Britain because we are kept warmer than countries of a similar latitude by gulf stream currents in the Atlantic.
- The ice caps at the Arctic and Antarctic (North and South Pole) may melt to some extent. Again nobody can really say how much this will affect sea levels. Melting sea ice will have little effect, but ice that melts over land masses may be much more significant.
- Higher temperatures make water expand, which some scientists think may have more of an effect on sea levels than ice melting.
- Rising sea levels will affect many low-lying countries in the world. Close to home, East Anglia and large areas of Holland will be flooded. Around the world there could be millions of people made homeless.
- Hurricanes form over water that is 26 °C or warmer, so it is likely that there will be more hurricanes.
- Higher temperatures could mean more evaporation, more clouds and a change in rainfall patterns.
- All over the world ecosystems depend on the climate. Changes will affect plant growth and, consequently, the pattern of agriculture.

The list of possible consequences to ecosystems is endless, but consider these:

1. In many places, temperature is a limiting factor. The higher the temperature, the greater the rate of photosynthesis and therefore the greater the yield. (The light-independent reaction of photosynthesis is controlled by enzymes and therefore is very temperature sensitive). So in some areas, agriculture might be more productive.
2. However, higher temperatures also lead to greater transpiration/evaporation rates, so yield will be reduced if water is in short supply.
3. Many crop varieties have been selectively bred so they are adapted to their current climate. If climates change, new varieties will be needed.
4. Many pest species are also adapted to the current climate. A change of climate may alter their range, so they spread. We might find malaria-carrying mosquitoes moving further north into Europe, or that some pests such as locusts complete their life cycle faster, greatly increasing their population size.

Examiners' Notes

This area of specification is an important 'interpret data' section, so you need to practise applying your knowledge of plant growth to questions that provide new and unfamiliar information.

5 Changes in vegetation will affect the rest of the food web. Plant productivity affects herbivores, which in turn affects the carnivores and the rest of the community.

6 The range of many organisms is limited by temperature. Increasing temperature may result in some tropical species extending their range, while native species might be out-competed by the newcomers.

The nitrogen cycle

Nitrogen is an essential component of several vital compounds including proteins, nucleic acids (RNA and DNA) and ATP. The nitrogen cycle is summarised below and is illustrated in Figs 20 and 21.

- Plants in general absorb nitrogen as nitrate (NO_3^-); a soluble ion that their roots can take in by active transport.
- Plants combine nitrate with the carbohydrate made in photosynthesis to make amino acids and nucleotides, the building blocks of proteins and nucleic acids.
- Nitrogen is passed up the food chain in these large organic molecules when animals eat the plants.
- Finally, all nitrogen ends up in non-living organic material. This could be dead leaves, dead bodies, faeces or urine (which contains nitrogen in urea or uric acid).
- The nitrogen-rich dead matter (**humus** or **detritus**) is broken down by decomposers, for example, bacteria and fungi. These organisms obtain their nutrients by **extracellular digestion** (secretion of enzymes and absorption of the soluble products, such as amino acids). This process is called **saprobiotic nutrition** (previously called **saprophytic decay**) and the end products are ammonium compounds (NH_4^+ ions).
- The last stage of saprobiotic nutrition (production of ammonium compounds from amino acids or other compounds) is **ammonification**.
- The ammonium ions are used by **nitrifying bacteria**. These organisms are examples of **chemoautotrophs**, and obtain their energy from oxidation reactions rather than from the Sun; the bacteria obtain their energy from the oxidation of the ammonium ions. This is called **nitrification**.

Essential Notes

Bacteria and fungi used to be referred to as **saprophytes** but now the term **saprobionts** is preferred. The terms mean the same thing – decomposers. The process of decay by bacteria releases ammonium compounds, and for this reason saprobiotic (saprophytic) nutrition is sometimes referred to as ammonification.

Examiners' Notes

Be careful to distinguish between the element nitrogen contained in compounds such as nitrates and protein, and nitrogen in the atmosphere. When referring to atmospheric nitrogen always write 'nitrogen gas'.

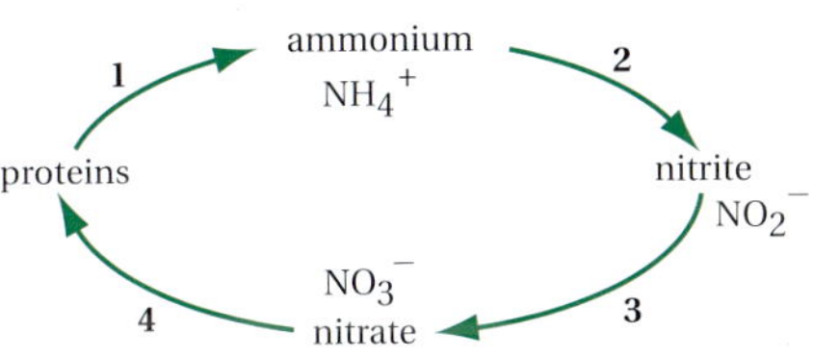

Key:

1	=	saprobiotic nutrition
2+3	=	nitrification
4	=	nitrogen passes up the food chain in organic molecules (mainly protein)

Fig 20
The nitrogen cycle

Examiners' Notes

Learn the basic nitrogen cycle, as shown in Fig 20, before you attempt to learn all its complications.

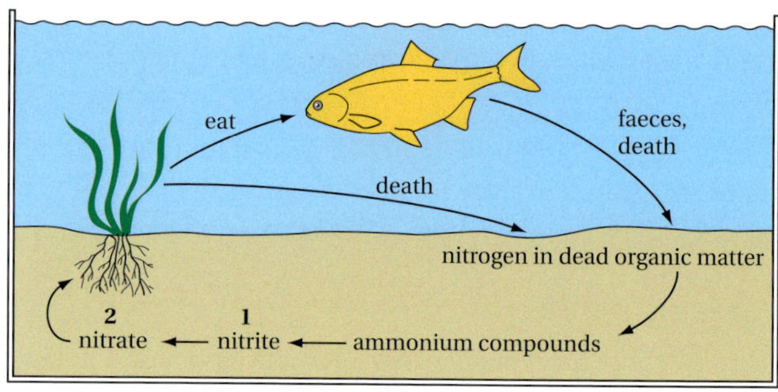

Fig 21
The nitrogen cycle in a fish tank

Nitrification

Nitrification is a two-stage process, as indicated by the numbers (1 and 2) in Fig 21, which shows the nitrogen cycle in a fish tank:

1. The ammonium ions are oxidised into nitrite (NO_2^-) ions by bacteria of the genus *Nitrosomonas.*
2. The nitrite is further oxidised into nitrate (NO_3^-) ions by bacteria of the genus *Nitrobacter.*

Finally, the soluble nitrate is absorbed by plants and the cycle repeats itself.

As well as the main cycle described above, you also need to know how nitrogen can be lost from the cycle to the atmosphere – **denitrification** – and how it can be regained from the atmosphere – **nitrogen fixation**.

Losing nitrogen from the atmosphere – denitrification

Denitrifying bacteria turn nitrate into nitrogen gas, thus losing it from the nitrogen cycle. These bacteria are anaerobes and so thrive in waterlogged soil, stagnant water and other oxygen-starved areas.

Gaining nitrogen from the atmosphere – nitrogen fixation

About 80% of the atmosphere is made up of nitrogen gas, but this is usually unavailable to living things. Molecules of N_2 gas have a triple bond ($N{\equiv}N$) that takes a lot of energy to break. When nitrogen gas is turned into soluble nitrogen ions, which are available to organisms, we say that nitrogen has been *fixed.* Nitrogen can be fixed during electrical storms, when the lightning provides enough energy to split the triple bond, so that the accompanying rain has dissolved nitrate in it. More reliably, however, bacteria can fix nitrogen. **Nitrogen-fixing bacteria**, mainly of the genus ***Rhizobium***, contain the enzyme **nitrogenase** that allows nitrogen gas to be fixed at low temperatures.

Nitrogen-fixing bacteria can be free-living in water or soil, but they can also occur in the roots of some plants such as **legumes** (for example, peas, beans and clover). These plants have a **mutualistic** relationship with *Rhizobium*; the plants get a supply of nitrate while the bacteria get some protection from predation and a supply of sugars. Legumes can thrive in nitrogen-poor soil. Therefore growing a legume crop and allowing it to decay in the soil is a natural way to improve soil fertility.

NB: The gene for the enzyme nitrogenase has now been isolated, which presents the possibility for it to be inserted into the genome of other plants. In theory, these too will be able to fix their own nitrogen and reduce the need for fertiliser.

Examiners' Notes

Make sure that you can distinguish between:

- nitrifying bacteria that turn ammonium into nitrates – 'the good guys'
- denitrifying bacteria that turn nitrates into nitrogen gas – 'the bad guys'
- nitrogen-fixing bacteria that turn nitrogen gas into ammonia – 'the angels'.

Eutrophication

Eutrophication means 'over-fertilisation' and it is a problem that affects waterways near agricultural land. Traditionally, farmers used organic fertiliser (such as cow dung) to maintain the fertility of the soil. Today, when many farms have no animals, this is increasingly difficult and farmers turn to inorganic fertilisers that can deliver exactly the right nutrients to the crop. This can cause problems because large amounts of the fertiliser can be washed away (**leached**) into rivers and lakes. This causes a nutrient build-up that results in the water being over-fertile. This process is summarised in Fig 22.

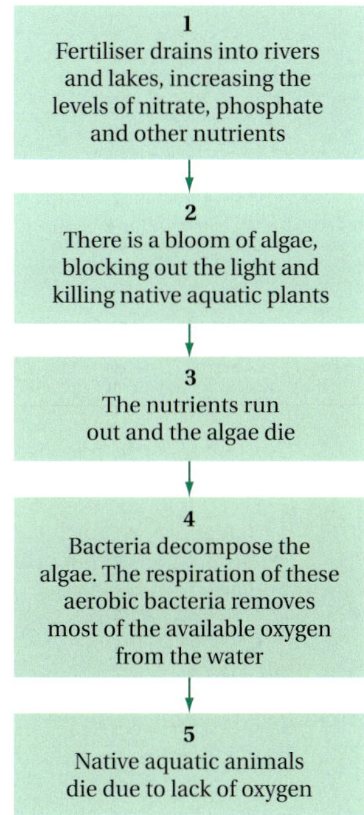

Fig 22
A summary of eutrophication

Examiners' Notes

Note that the algae themselves do not reduce the oxygen content. They photosynthesise, so actually increase the oxygen levels for a short while.

3.4.7 Ecosystems are dynamic systems, usually moving from colonisation to climax communities in the process of succession.

The development of an ecosystem

Before deforestation in the last thousand years or so, the UK was largely covered with **deciduous forest**: broad-leafed species that shed their leaves in the autumn. The dominant species were oak, ash, beech and birch, among others. This is the **climax community** that develops in our temperate climate. Bears, lynxes and wolves roamed the countryside; the human population was small and its influence was negligible. Now there is very little 'natural' forest left – virtually none in England – the forests that do exist were planted by humans.

So how did the forest develop in the first place? Ecosystems develop by the processes of **colonisation** and **succession** until the climax community is established (Fig 23).

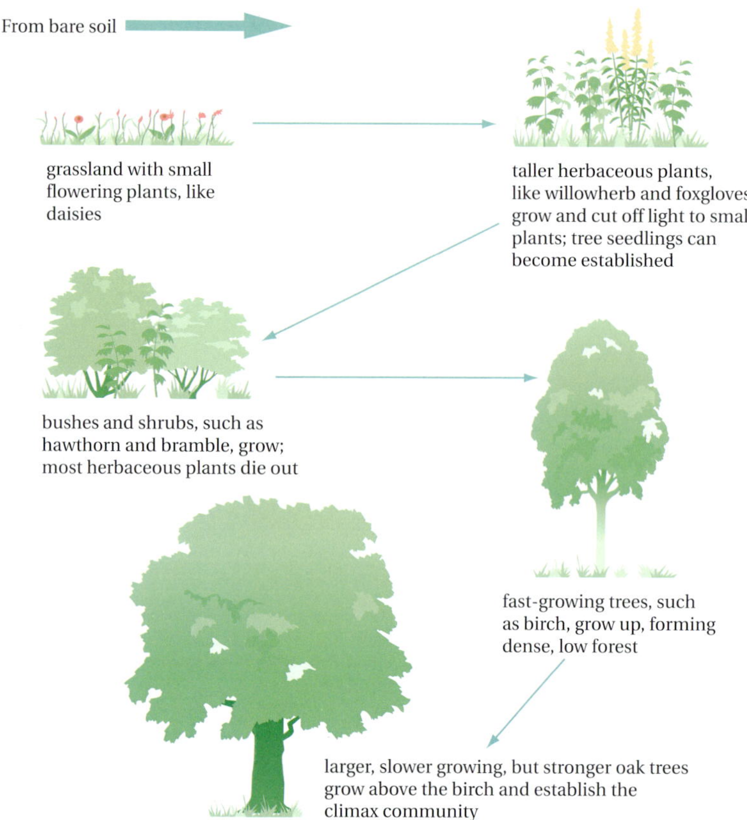

Fig 23
Succession that occurs on a patch of bare ground, assuming little or no grazing. Deciduous (or broadleaf) trees lose their leaves in winter; in contrast some trees are evergreen, for example, the pine trees that dominate the forests of colder latitudes; herbaceous plants have no woody tissue, in contrast to shrubs and trees

The process of succession normally takes place over decades or centuries, but there are two common ways of studying the process:

- Clear a patch of ground and watch what happens to the bare soil.
- Study a sand dune system near a beach and observe the changes that occur as you move inland.

Clearing a patch of ground

Remove all the plants, fence it off from grazing animals, and observe the changes. The problem is that it takes at least 50 years, but eventually the forest is re-established. There are two conditions that need to be met if the ecosystem is to develop as normal:

- The soil must initially have a relatively low humus content.
- There must be no grazing animals that can get onto the plot.

The main stages that you would observe as the ecosystem develops are:

- **Colonisation** – the bare soil is colonised by what we might think of as weeds – herbaceous plants that can grow in relatively nutrient-poor soil. They have a rapid life cycle, then die back and increase the humus content of the soil. Typical coloniser species are grasses, daisies, dandelions and clover.
- **Succession** – this occurs because the colonisers change the habitat. Once the colonisers have improved the quality of the soil, by dying and rotting, more species can grow. The **diversity** of herbaceous plants increases greatly

Examiners' Notes

In examination questions, avoid being anthropomorphic, i.e. giving human emotions to non-human organisms; for example, trees do not '*like*' light, woodlice are not '*happy*' in rotting vegetation and plants don't '*know*' when it's time to flower.

Examiners' Notes

It is a good idea to learn one example of succession in detail, and make sure that you can name some plant species at each stage, rather than talking vaguely about 'trees and shrubs', etc.

as the abiotic environment becomes more favourable. Taller plants cut off the light and so out-compete the shorter plants. The greater diversity of plants attracts more insects, which attract birds and small mammals. At this stage grazing can have a marked effect and prevent any further succession. Many plants, including tree saplings, have their growing points (meristems) at the top of the stem, and herbivores such as rabbits and sheep prevent any further growth. In contrast, grasses grow from the base of the stem, so they thrive despite constant grazing.

- **Establishment of the climax community** – in the next phase, small woody plants – shrubs such as hawthorn and bramble – begin to dominate. In turn, these are out-competed for light by fast-growing tree species such as birch that form a low, dense forest. Eventually the large, but slow-growing, trees – notably the oak – begin to dominate until the climax community is established and there is no further succession.

The climax community that develops depends on the climate. The process outlined above will not happen, for example, on exposed hillsides where the soil is too thin or the wind too harsh. Other examples of climax communities around the world include rainforest, cloud forest, tundra, grassland and coniferous forest.

Studying a sand dune system

Sand dunes are useful areas to study because you can observe colonisation and succession without having to wait 50 years. You can see some of the changes associated with the development of ecosystems as you simply walk inland (Fig 24). Near the sea the dunes are at their youngest – wind tends to pile up the sand and the profile changes from year to year.

Sand is a very difficult medium for plants: water and nutrients drain straight through, and the constant shifting makes it very difficult for the roots to anchor the plant. However some **pioneer species**, notably sand couch grass, lyme grass and marram grass in the UK, have a dense root system that binds the sand together. This holds water and humus particles and makes the whole dune more permanent. Sometimes the marram grass is the pioneer species, and sometimes the sand couch grass and lyme grass are the pioneers due to their greater

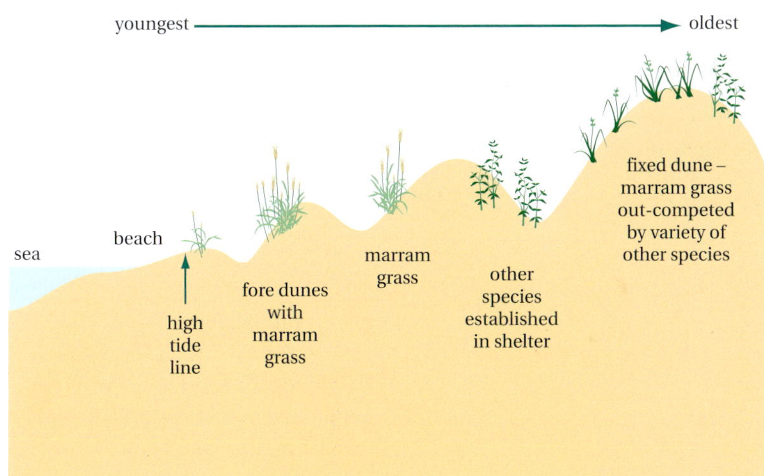

Fig 24
Sand dune succession

tolerance to salt. In most dunes, marram grass is the key species because it makes the environment less hostile, so that other plants such as ragwort, willow and grasses can take over. As you move inland the sand gets darker because the humus content increases and so does the species diversity.

Succession from bare rock

Although our planet seems pretty crowded, there are still a few places where ecosystems can develop from scratch. A good example is the areas created by volcanic lava. As the lava cools it creates a barren, hostile environment but even this can be colonised by organisms such as bacteria, fungi, algae and lichens. (You can often observe the same process beginning on the tiles on your roof.) As the rock weathers, soil particles gather in cracks so that mosses and other small, shallow-rooted plants can get a hold. As these plants spread they create yet more soil, which increases in both depth and nutrient content. Succession then takes place in the manner described above.

Essential Notes

Lichens consist of algae growing inside a fungus. The fungus provides anchorage and protection from drying out. In turn, the algae can photosynthesise and give the fungus organic compounds that would otherwise be unavailable from bare rock.

Species diversity

The numbers of different species that form the community of an ecosystem can vary greatly. Environments such as a coral reef or a rainforest show a **high species diversity** because conditions are generally favourable and stable. In these situations biotic factors – those due to other organisms – dominate organisms' lives. In contrast, in harsh environments such as the arctic or desert regions, there is a **low species diversity** and abiotic factors, such as temperature and water availability, dominate.

Managing succession

An understanding of how to control ecological succession is important in a variety of situations:

- When growing trees and other plants in a **sustainable** way. Different trees grow at different rates, and appear at different stages of a succession. By harvesting fast-growing trees at the correct size and age, foresters can maximise sustainable wood production. This approach can also maximise **biodiversity**, which is often at its greatest during a succession rather than at the end. This increases the number of potentially useful plants and provides more habitats for animals. The basic stages of a succession were shown in Fig 23. It can be seen that it will take a lot longer to harvest slow-growing trees such as oak, than it will to harvest the faster-growing trees such as birch.
- In some cases, succession is controlled because of its aesthetic value. For example, the appearance of the Lake District with its rolling landscapes is maintained largely by sheep farming – because grazing prevents succession back to woodland. Sheep farming is usually unprofitable but tourism is not, so farmers are actually paid subsidies to maintain the appearance of the countryside.
- Ponds are not usually permanent features. Over time, they 'silt up' with dead leaves and soil, etc., so that they dry out. Then the usual land succession takes over. However, human activity often reduces the number of new ponds that can develop, so old ponds can be maintained by dredging, thus preventing them from silting up. In this way, a valuable habitat is saved and biodiversity is maintained.

3.4.8 Genetic variation within a species and geographic isolation lead to the accumulation of different genetic information in populations and the potential formation of new species.

This section builds on the AS Unit 2: *The Variety of Living Organisms*. You need to be familiar with the structure of DNA and the basics of cell division (**meiosis** and **mitosis**). In particular, *independent assortment* in meiosis is fundamental to an understanding of genetics.

Definitions

Genetics: some key definitions

Genotype – *the genetic constitution of an organism. When we study the inheritance of one gene we use the term genotype to describe the combination of* **alleles** *the individual possesses. This is written as, for example, BB or Bb.*

Phenotype – *the observable features of an organism. The phenotype results from a combination of the genes the organism inherits and the environmental factors. For example, we are all born with a certain combination of tallness genes so we have the potential to grow to a particular height. However, we will not reach this height without an adequate diet.*

Put simply: **Phenotype = genotype + environment**

You will also need to know the meaning of the following terms:

Gene – *a length of DNA that codes for the manufacture of a particular protein or polypeptide. In this section, think of it as a piece of DNA that codes for a certain feature. Humans, like all other animals, are* **diploid** *organisms – we have two copies of each gene.*

Chromosome – *one long DNA molecule with genes dotted along its length (Fig 25).*

Locus – *the position of a gene on a chromosome.*

Allele – *an alternative form of a gene. Genes often have two or more alleles; for example, a flower colour gene could have two alleles, such as red and white.*

Dominant – *the allele that, if present, is shown in the phenotype; for example, the allele B could cause black fur in mice.*

Recessive – *the allele that is only expressed in the absence of the dominant version; for example, the allele b could code for brown fur in mice.*

Homozygous – *when both alleles are the same: BB or bb.*

Heterozygous – *when a pair of alleles is different: Bb.*

When an egg is fertilised a single set of chromosomes (Fig 25 overleaf) from the male matches with a single set from the female, so there are two sets of chromosomes. In these diploid cells there are two copies of each gene - one allele from the mother and one from the father.

Fig 25
Chromosomes contain many genes that are lengths of DNA that code for a particular protein

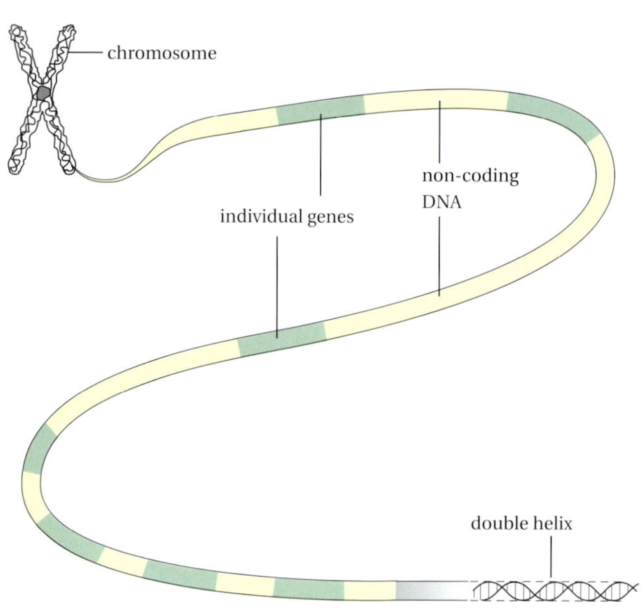

Mendel's genetics – monohybrid and dihybrid inheritance

The basic rules of inheritance were worked out in the 1860s by Gregor Mendel. However, his work remained largely unappreciated until around 1900, when the process of meiosis was first observed and understood. Meiosis (see *Collins Student Support Materials Unit 2: The Variety of Living Organisms*) went a long way to explaining Mendel's observations.

Table 6
Possible genotypes for mice with coat colour B and b

Genotype	Phenotype
BB (homozygous dominant)	Black
Bb (heterozygous)	Black
bb (homozygous recessive)	Brown

Monohybrid inheritance

This refers to the inheritance of a single gene with two alleles. For example, coat colour in a particular strain of mice is determined by one coat colour gene with two alleles, B (for black fur) and b (for brown fur).

There are three possible genotypes for these mice, but only two phenotypes, as shown in Table 6.

Note that black is dominant because if the allele B is present, it is shown in the phenotype. The b allele is recessive, and is only expressed in the absence of the other allele. Fig 26 shows the basics of a **monohybrid cross**. If a Bb mouse mates with a Bb mouse, there is a three-to-one ratio in the phenotypes of the offspring.

In genetics we deal with large numbers of gametes, so the ratios are averages. It's not like the draw for the FA Cup, where you know what's been drawn and therefore what's left in the bag. If the chances of a child being born with genotype bb is 1 in 4, and the couple have three children of genotype BB, Bb and Bb, then the chance that the fourth will be bb is still 1 in 4.

Examiners' Notes

Every year candidates lose marks by writing ratios in the wrong way. A ratio of 1:3 is not the same as 1 in 3; the ratio of 1:3 is the same as 1 in 4, and can also be expressed as 0.25 or 25%.

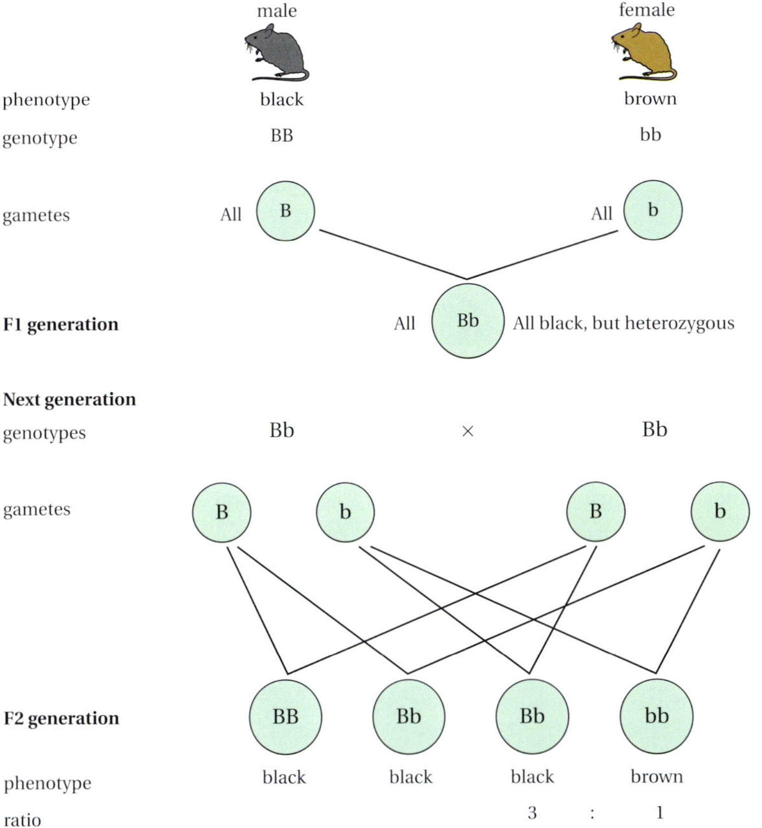

Fig 26
Monohybrid inheritance in mice. Pure breeding black and brown mice are homozygous (BB and bb). When bred, the mice born in the first litter are all black and are heterozygous (Bb); if these mice are then bred together, the litters have black and brown mice in a ratio of 3:1

Codominance

There are cases where alleles are **codominant**, so if they are present they are both expressed in the phenotype. A classic example is the human ABO blood group system (which is also an example of a **multiple allele**) where there are more than two alleles of the same gene. Some genes have many different alleles.

The ABO system is controlled by one gene, I, with three alleles, I^A, I^B and I^O. I^A and I^B are codominant over I^O.

- Allele I^A codes for A proteins on the red cells.
- Allele I^B codes for B proteins on the red cells.
- Allele I^O codes for no relevant proteins on the red cells.

This gives us the table of genotypes and phenotypes shown in Table 7.

Table 7
The genotypes of blood groups

Genotype	Phenotype (i.e. blood group) and explanation
I^O, I^O	O – I^O is recessive
I^A I^O or I^A I^A	A – I^A is dominant over I^O
I^B I^O or I^B I^B	B – I^B is dominant over I^O
I^A I^B	AB – I^A and I^B are codominant

Dihybrid crosses

A **dihybrid cross** involves two separate genes, each with two alleles, at the same time. It is assumed that the genes are located on different chromosomes so that they can be separated by meiosis.

The vital steps in working out the products of a dihybrid cross (Fig 27) are:

1 Parents' genotypes.

2 Gamete formation – this is where you can apply your knowledge of meiosis. One allele from each pair can go into the gamete, but by independent assortment it can pass into the gamete with any allele from the other pair. Thus a genotype of AaBb can give gametes of AB, Ab, aB and ab.

3 By random fertilisation any male gamete can combine with any female gamete.

4 With dihybrid inheritance there can be up to 16 different genotypes, so it is important to organise the results using a **Punnett square**.

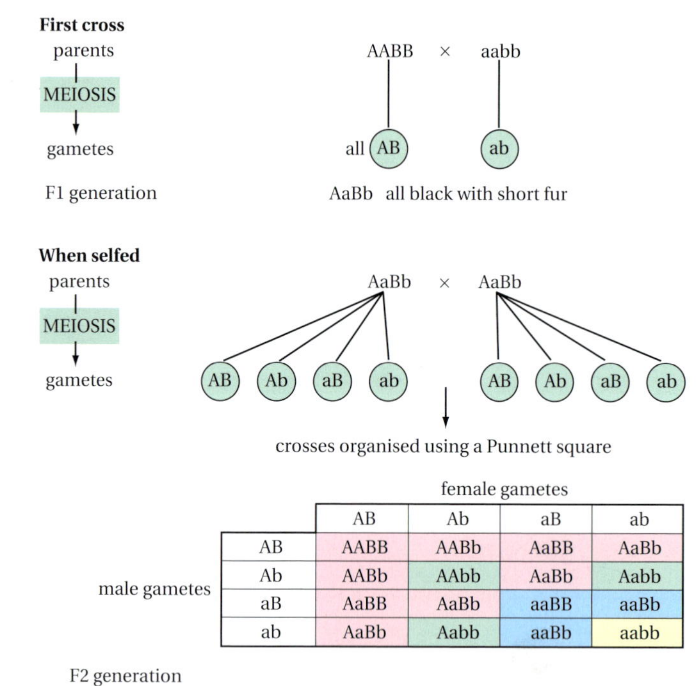

	AB	Ab	aB	ab
AB	AABB	AABb	AaBB	AaBb
Ab	AABb	AAbb	AaBb	Aabb
aB	AaBB	AaBb	aaBB	aaBb
ab	AaBb	Aabb	aaBb	aabb

F2 generation

Black, short fur	Black, long fur	brown, short fur	Brown, long fur
9	3	3	1

Fig 27
An example of dihybrid inheritance – a cross involving two genes. We have already looked at the mouse coat colour gene, where the allele B for black hair is dominant to the allele b for brown hair. A second gene controls hair length. Allele A for short hair is dominant to allele a for long hair. A cross between two pure-breeding mice, one black with short hair and the other brown with long hair, will produce a litter of all-black mice. However, these mice are all heterozygous for both genes. If these mice are bred together to produce a second generation (F2), all possible combinations are produced in the classic ratio of 9:3:3:1

Sex linkage

Sex-linked inheritance concerns genes found on the sex chromosomes, X and Y. The X chromosome is larger and contains more genes, so most examples of sex linkage concern X-linked genes. Y-linked traits are known, but are rare.

With sex-linked inheritance, the normal pattern is affected by the key fact: *males only have one X chromosome; so have only one copy of a sex-linked allele.*

Haemophilia is an example of a sex-linked disease. This disease is caused by a faulty allele that fails to make the correct blood clotting protein, **factor VIII**. There are two alleles: the normal allele, H and the faulty allele, h. In sex linkage we show the chromosomes as well as the alleles.

There are three possible genotypes for females:

- X^HX^H = a healthy female; normal alleles on both X chromosomes
- X^HX^h = a healthy female, but a carrier of the haemophilia allele
- X^hX^h = a haemophiliac (very rare in females).

Males have only two possible genotypes:

- X^HY^- = a healthy male; there is no allele on the Y chromosome
- X^hY^- = a haemophiliac; he has no healthy gene on the Y chromosome to mask the effects of the faulty gene he has inherited

The Hardy-Weinberg principle

In this section we look at populations rather than individuals. A **Mendelian population** or **deme** is a population of organisms of one species that can actively interbreed with one another and share a distinct **gene pool**.

Imagine a gene for coat colour in a particular population of rats. Allele A is the dominant allele, producing brown fur. Recessive allele a produces albino rats. Rats of genotype AA and Aa are brown, while aa rats are albinos (they have white fur due to an absence of pigment).

It's easy to find out how many rats are genotype aa because we can see them. But what if we want to know how many are Aa? They look just the same as AA rats. We can use the **Hardy-Weinberg** equation to work out how many rats are carrying the recessive allele.

- If we use the letter p to denote the frequency of the A allele.
- And q for the frequency of the a allele.
- We know that $p + q = 1$ (1 is the same as 100%. There are only two possible alleles so when added up they must equal 100%.)

Each individual has two alleles.

- The frequency of AA rats is p^2.
- The frequency of Aa rats is $2pq$ (because the rats can be aA or Aa).
- The frequency of aa rats is q^2.

And as these are the only three possibilities:

$$p^2 + 2pq + q^2 = 1$$

In words, this formula says that the frequency of AA rats, when added to those that are Aa, aA, or aa, must equal 100%.

Essential Notes

A key point here is that, with sex-linked inheritance, males can't be carriers.

Worked Example

If 4% of the rats are albino:

(a) What is the frequency of the a allele?

(b) What proportion of the rat population is heterozygous, i.e. has the genotype Aa and so carries the a allele?

Answers:

(a) If 4% of the rats are albino, and therefore aa, we know that $q^2 = 0.04$.

Then the frequency of the a allele is $\sqrt{0.04}$, which is 0.2 or 20%.

We know that $p + q = 1$, so the frequency of the A allele is 0.8, or 80%.

(b) The question is asking 'how many rats are Aa?', which is $2pq$.
$2pq = 2 \times 0.8 \times 0.2 = 0.32$ or 32%.

Overall, from the observation that 4% of the rats are albino the Hardy-Weinberg equation tells us that 32% of the rats are Aa and 64% are AA.

In population genetics the Hardy-Weinberg principle states that the genotype frequencies in a population *remain constant* or are in equilibrium from generation to generation unless *specific disturbing influences* are introduced. These include the following:

- **Non-random mating** – for example, if brown rats tended to choose other brown rats as mates, or albinos only mated with other albinos.
- **Mutations** – so that new alleles were created, more than just A and a.
- **Selection** – where one coat colour gives the individual an advantage over other individuals with the other coat colour.
- **Limited population size** – when numbers are small, chance becomes significant.
- **Gene flow** – immigration or emigration from the population.

Selection

Variation between individuals is vital because it helps populations to survive unfavourable conditions, and in practice this is nearly always the case. There is competition for resources: food, light, water, nutrients, etc. In addition, organisms have to survive drought, cold, heat and disease.

Natural selection is often called *survival of the fittest*, but in a biological context, *fitness* means **reproductive success**. Fit organisms will pass on more of their genes to the next generation than will less fit organisms.

Definition

Selection: a key definition

Natural selection is a process in which those organisms whose alleles or allele combinations give them a selective advantage are more likely to survive, reproduce and pass on their alleles to the next generation.

A good example of natural selection in action is that of antibiotic resistance in bacteria. What the media call 'superbugs' are bacteria that resist treatment with antibiotics. Some strains of bacteria are resistant to virtually all of the antibiotics at our disposal and they represent a growing threat to our health.

Initially, the resistance probably arose as a result of a mutation. A single gene in a bacterium mutated into a new allele, which produced a protein that in some way made the bacterium resistant to the antibiotic.

It is important to note that the antibiotic *did not cause* the mutation, which must have already existed in the population; or it occurred in the right place at the right time, to give the individual a selective advantage. In areas of antibiotic use, the resistant bacteria multiplied, while the susceptible bacteria were killed. In this way, the *frequency of the resistance allele* increased over time.

Natural selection is often thought of as just a mechanism for change, but it can also be a force that keeps things stable. There are three types of selection. (These are shown in Fig 28 and described overleaf.)

Examiners' Notes

Populations (refer back to page 10) contain individuals with many different mutations that have accumulated over time. A change in the environment – such as a new antibiotic – will make a particular mutation relevant. Students often lose marks by stating that organisms mutate in *response* to the change.

Fig 28
Different types of selection: the shaded areas in the top row of graphs show individuals with a selective advantage

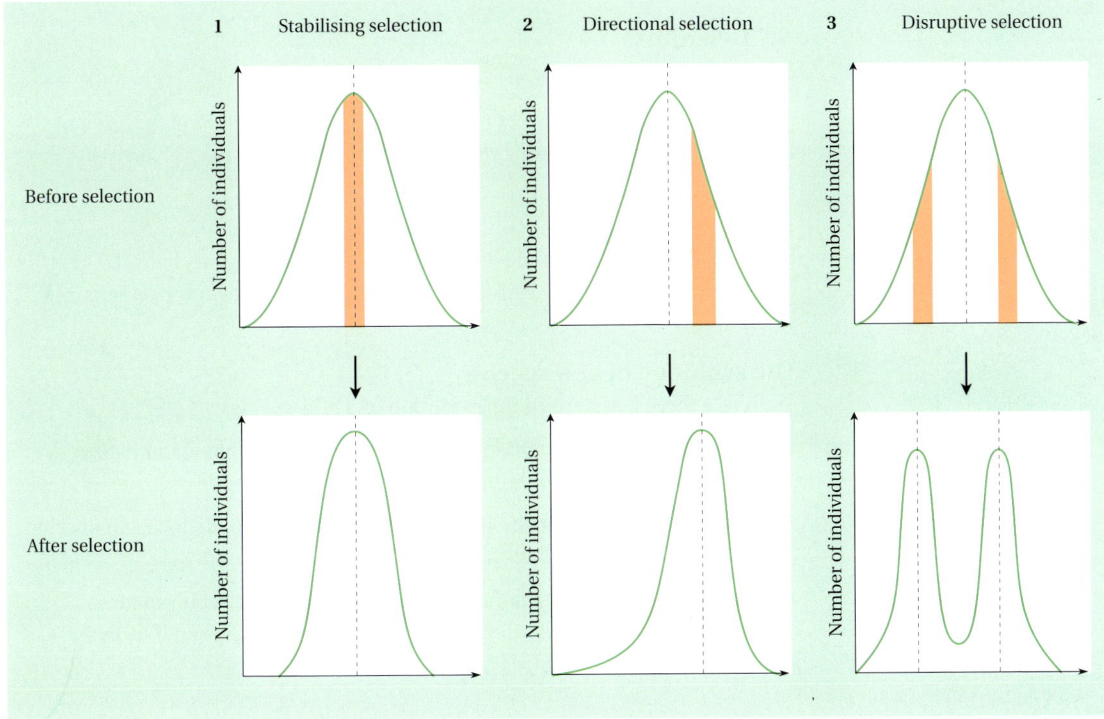

1 **Stabilising selection** – this can lead to a standardising of organisms by selecting against extremes, especially in a stable environment. Birth weight of mammals is a good example. An abnormally large baby may make the pregnant female too heavy and therefore unable to hunt or avoid predators; it may also cause problems to both mother and baby during birth. Conversely, a very small baby may not be strong enough to walk/run/keep pace with its parents, and might therefore be vulnerable to predators.

2 **Directional selection** – this tends to occur following some environmental change which causes selection pressure. Organisms with a particular extreme of phenotype may have a selective advantage. The giraffe is a classic example: probably following a drought only the tallest animals could reach vegetation and so got enough food to survive and pass on their alleles/allele combinations to the next generation.

3 **Disruptive selection** – this is selection against the middle, favouring the individuals with extremes of phenotype. This was seen in certain species of finches on the Galapagos Islands. Starting with a common ancestor that showed a range in beak size, those with smaller beaks had an advantage when catching insects, while those with larger, stronger beaks had an advantage when breaking open seeds. Those with a phenotype on the mid range were out-competed by those at the extremes.

Speciation

Defining a species is difficult because new species evolve from existing species and the process is gradual. However, for examination purposes a working definition of a species is as follows:

Definition

Species: a key definition

A population or group of populations of similar individuals that can mate and produce fertile offspring.

The classic example of this concept is the horse and the donkey. A male horse can mate with a female donkey to produce a mule. This hybrid, however, is sterile and so the horse and the donkey are regarded as separate species.

The evolution of new species

There are three key steps to the evolution of a new species:

1 **Isolation** – part of a population becomes isolated so that it *cannot breed* with the rest of the population.

2 **Natural selection** – acts in different ways according to the local situation to *change the frequency of alleles* and, eventually, the phenotypes.

3 **Speciation** – over the generations, genetic differences accumulate so that the different populations are unable to interbreed, even if brought together again.

Fig 29
An example of speciation

Steps in the evolution of a new species:

1. A population of rabbits on an island is split by a new river, so the two populations are isolated
2. Natural selection acts in different ways on the two populations: in the rocky area, for example, the rabbits can't burrow, so are more vulnerable to predators; natural selection therefore favours those with the keenest senses, best reflexes and greatest speed, and they evolve into hare-like creatures
3. Over the generations, genetic differences accumulate so that even if the two populations come into contact, they are not able to breed successfully

Isolating mechanisms

We have seen that a key feature of speciation is isolation: where one population is prevented from interbreeding with another population. Isolating mechanisms are generally divided into two categories, although there are cases that don't fall neatly into either one:

- **Allopatric speciation** – the populations are physically separated and so cannot interbreed. Barriers include water, mountain ranges or, in recent times, man-made obstacles such as farms or even roads. This type of speciation is easy to explain and examples are common.
- **Sympatric speciation** – the populations are not physically separated, but still do not interbreed and evolve along different lines. Sympatric isolating mechanisms are nowhere near as clear-cut and easy to explain as allopatric isolating mechanisms. Examples are rare and arguments between scientists are common.

Examiners' Notes

Domestic dogs come in all shapes and sizes, but a Jack Russell and a poodle can mate to produce perfectly fertile crossbreeds. We say that the poodle and the Jack Russell are different *breeds*, but not different species.

How Science Works

In science, we make advances by a combination of two processes:

1 **Observation** – we look at the world and say 'Could it possibly be...?' and then come up with testable ideas. We call these *hypotheses*.

2 **Experimentation** – we gather evidence and analyse the data to draw reliable conclusions. Sometimes we gather support for our hypothesis, and sometimes we disprove it. Importantly, we never, ever, prove anything. So don't write this in your conclusions.

From your practical investigations in biology you will already be familiar with many of the basic principles used by scientists in their research. The rules that follow should be applied to any scientific investigation.

Testing a hypothesis

Fig H1 summarises the stages in scientific research.

Progress in science is made when a hypothesis is tested by an experiment. Contrary to popular belief, scientists do not just do experiments to see what happens. Fun though it might be, they don't just mix chemicals together and watch the results.

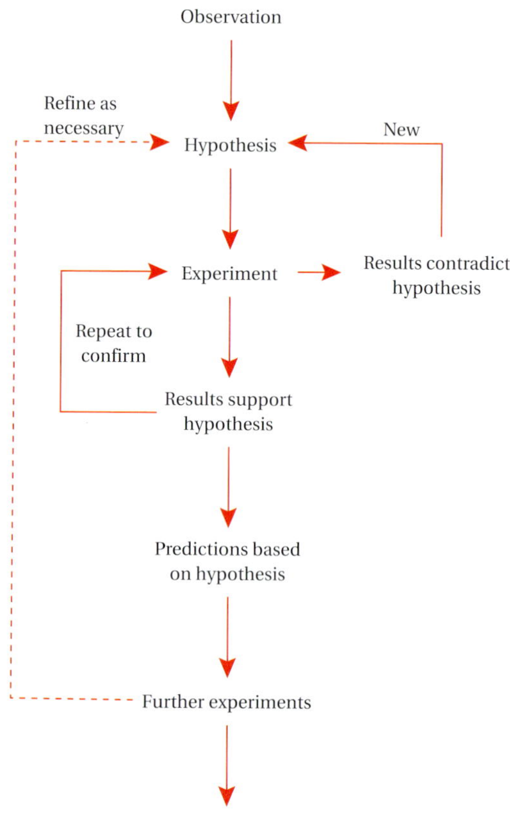

Fig H1
The stages of scientific research

An experiment must be designed to test one possible explanation of an observation. The definition of a good hypothesis is that an experiment can either support or disprove it. Strictly speaking, experiments can never prove that a hypothesis is absolutely, definitely correct. There is always the possibility that another explanation, one that no one has thought of, could fit the evidence equally well. However, an experiment can prove that a hypothesis is definitely wrong.

Unfortunately, people are often tempted to bypass the testing stage and go straight to an explanation without obtaining any experimental evidence. Some people seem to be uncomfortable when they are unable to find an explanation for a phenomenon. Even scientists have a tendency to be biased towards finding evidence to support their hypothesis.

As a student, you may well have done an investigation during which you were disappointed to get results that disproved your hypothesis. Or maybe your results were not what you expected. When this happens students often suggest that their experiment has 'gone wrong'; but, in scientific research, negative results are just as important as positive results.

When does a hypothesis become an accepted theory?

A hypothesis only becomes accepted theory when it has been thoroughly tested. The hypothesis may suggest predictions that, in turn, can be tested by further experiments and observations. Other scientists try to think of alternative interpretations of the results. It is normal practice for one scientist to be critical of another's published results. To ensure that published work is of sufficiently high quality, journals practise 'peer review' – a submitted paper is reviewed by two or three other experts in the field to make sure the experiments have been carried out well, and that the way the results have been interpreted is reasonable.

Essential Notes

The term 'expert in the field' is used to describe a scientist with experience and a great deal of knowledge in a particular area of science. In peer review, the scientists reviewing their colleagues' (peers') research, need to work in the same area of science, to be able to give a reliable and useful opinion of its quality.

It should also be possible to repeat an experiment and get the same results. Only after many confirmatory experiments is it likely that a new idea will be accepted. For example, for many years it was thought that cell membranes had a structure rather like a sandwich with protein 'bread' and phospholipid 'filling'. After many experiments this hypothesis was shown to be false, and it has now been replaced by the fluid-mosaic explanation described in this unit. This idea is now so well supported that it is described as *the theory of plasma membrane structure.*

Once a hypothesis is supported in this way – by many experimental results and observations – it may be accepted as the best explanation of an observation.

A theory is, therefore, a well-established hypothesis that is supported by a substantial body of evidence. The theory of natural selection, for example, is based on huge numbers of observations, predictions and experiments that support the underlying hypothesis.

Designing an investigation

Suppose you are asked to design an experiment to investigate the effect of temperature on the rate of reaction of an enzyme such as catalase.

Your hypothesis could be:

Temperature has an effect on the rate of enzyme-controlled reactions.

Variables and controls

Catalase breaks down hydrogen peroxide to water and oxygen. To investigate the effect of temperature on the reaction, you could set up water baths at a range of temperatures, mix the catalase and hydrogen peroxide and measure the amount of oxygen released at each temperature.

There are, of course, practical difficulties to be overcome, such as collecting the oxygen without letting any escape, but in principle the experiment is quite simple. The key to this and all similar experiments is that you do three things:

- Select and set up a range of different values for the factor whose effect you are testing, in this case temperature. This factor is the *independent variable*.
- Measure the change in the factor that you are testing, in this case the rate of oxygen production. This factor is the *dependent variable*.
- Keep all other factors, such as enzyme and hydrogen peroxide concentrations, the same. These are the *controlled variables*.

Examiners' Notes

- The *independent* variable is the one the experimenter changes. The *dependent* variable is the one the experimenter measures.
- All other possible variables are kept constant.

Including a control experiment

One other precaution is to carry out a *control* experiment. This is not the same as keeping other variables constant. Its purpose is to ensure that changes made to the independent variable have not in themselves changed any other factor, and that the results really are due to the factor being tested.

For example, in the enzyme investigation featured above, how do we know that it is the enzyme that is breaking down the substrate and not simply the effect of temperature, or some other chemical in the enzyme solution? To answer this question, we must do a control experiment in which the enzyme is first boiled (to denature it), or left out altogether. If no oxygen is produced, we have shown that it really was the enzyme that was catalysing the reaction, not another factor.

Another example of a control can be taken from the common practical to test how effectively different antiseptics kill bacteria. Paper discs soaked in the antiseptic might be placed on a bacterial lawn in a Petri dish (see Fig H2).

In this experiment four of the discs were soaked in different antiseptics. The fifth disc was the control. The control disc should not be just a plain paper disc, but a disc that has also been soaked in sterile water, or whatever solvent was used in the antiseptics. This would show that the results obtained were really due to the antiseptics and not, perhaps, to something that could dissolve from the paper disc.

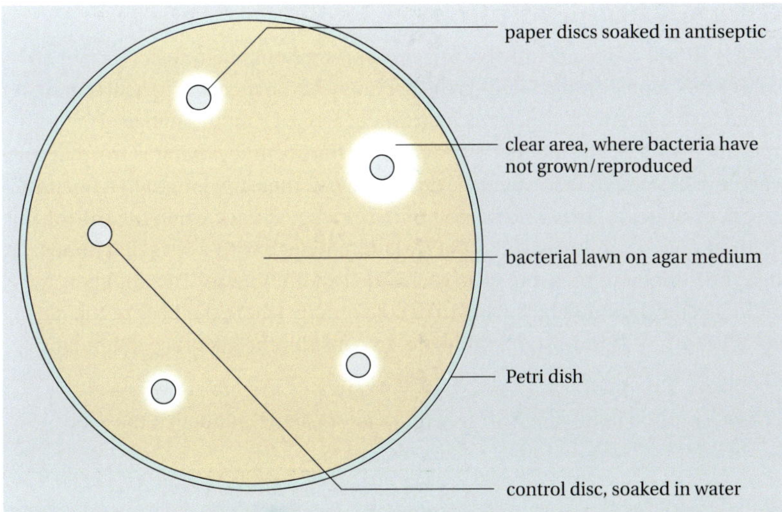

Fig H2
The effect of antiseptic discs on bacteria growth

What happens if it is not possible to control all the variables?

Sometimes it is not easy to ensure that all the major factors have been controlled. In an experiment on enzyme activity, controlling all the variables, apart from the independent variable you want to test, is quite straightforward. However, when experimenting with living organisms, investigations are rarely so simple because living things, themselves, are so variable.

If you are measuring the response of an animal to a stimulus, such as a woodlouse to light, you can never be sure that every single woodlouse will respond in the same way. Although most woodlice will move away from light, some might not. Even the simplest organisms respond to many stimuli. A living thing may behave untypically according to how well fed it is, its age, the time of day, its sexual maturity, or just because it is genetically different from most.

The only way to deal with this uncontrollable variability is to *repeat* an experiment several times and to use a large number of different organisms. Even in an experiment like the one using catalase it would be necessary to repeat the procedure several times for each value of the independent variable. For example, you would make several measurements of the rate of oxygen production at 35 °C, and several at 40 °C. Each repeat is called a *replicate*. Repetition increases the *reliability* of the results, and this increases the likelihood of being able to draw *valid* conclusions.

For students, there is nearly always a limit to the number of times an experiment can be repeated. Even researchers have time and resource constraints, and it is necessary to use judgement about the likely reliability of a set of data. If the results from replicates are all very similar, it is more likely that the results are reliable.

Accuracy and limitations

There is a limit to the accuracy of any measurement made in the course of an experiment. One limiting factor is the accuracy of the measuring instrument. A second is the care with which the instrument is used. But in biological experiments there is often a practical limit to the accuracy that it is worth trying to achieve. Although instruments exist which can measure length to a fraction of a micrometre, there would be no point in such accuracy when measuring tail length in mice in an investigation of variation. In fact, with a wriggling mouse, it might be difficult to measure even to the nearest millimetre. This difficulty would be compounded by having to decide exactly where the base of the tail actually starts. It is important, therefore, to consider the accuracy that might reasonably be expected from a set of data.

Accuracy is often confused with *reliability*. Consider the data in Table H1.

Table H1
Comparing loss in mass of leaves from two different types of plants

Leaf	Loss in mass over 24 hours/gram	
	Plant A	**Plant B**
1	1.03	0.28
2	0.96	0.72
3	0.89	0.74
4	1.05	0.69
5	0.94	0.64
Mean	0.968	0.64

Taking measurements from several specimens increases the reliability of the results, but it does not make them more accurate. For Plant A, all the results are reasonably similar, which suggests that the value for the mean is probably quite reliable. However, if another five leaves were measured, it is highly unlikely that exactly the same mean would be obtained.

The mean is given to three significant figures, but the results only to two significant figures. It is clearly absurd to give a value for the mean that is more precise than the accuracy of the measurements. Calculators give answers to many decimal places, but judgement has to be used about the number of significant figures that can sensibly be given in data for means, or other calculations that are derived by manipulating raw data.

The mean for Plant B looks unreliable. The result for Leaf 1 is very different from all of the others, so the mean comes well below all the other results. It may be that this result was a mistaken reading of the balance. On the other hand, the anomaly may have been because the leaf was atypical: it may have been much smaller, with fewer stomata than normal, or half-dead, for example. Without information about the original masses from which the losses were calculated it is impossible to guess. Expressing the results as a percentage loss rather than as total loss would make comparison more reasonable.

Associations and correlations: What affects what?

Many biological investigations depend on a combination of observation and data analysis rather than on actual experiments. This is because it is often not practical to carry out proper controlled experiments with living organisms in the field. There are two reasons for this:

- Logistical reasons – the complexity of interrelationships between organisms and the environment makes it virtually impossible.
- Ethical reasons – it is, for example, not ethical to remove the whole population of one species in an ecosystem in order to find the effect on the food web. Similarly you can't experiment on the effects of smoking by taking two groups of people and making one group smoke and the other group not, while keeping all other factors the same.

Investigators, therefore, have to look for associations that occur in the normal course of events. However, care needs to be taken when drawing conclusions. The number of fish in a lake affected by acid rain or some other pollutant may decline, but this does not necessarily mean that the pollution has caused the decline, or even that the two are connected. Further investigations could look for data on natural populations of particular fish species in water of different acidity. It would also be possible to carry out laboratory experiments to determine fish survival rates in water of different acidity. Results might well show that the lower the pH, the lower the survival rate. In this case there would be a correlation between pH and fish survival. This would still not prove that the decline in fish numbers in the lake was actually caused by the acidity.

If you counted, say, the number of nightclubs and pubs and the number of churches in several towns and cities and then plotted a graph of one against the other, you would almost certainly find a correlation. But this would obviously not prove that churches cause nightclubs and pubs to be built, or the other way round. The correlation is likely to be the result of a completely separate factor, probably the size of the town or city.

Similarly the decline in fish numbers might be due to some other factor, which might or might not be due to acidity. There could be an indirect association, caused by the effect of acidity on the food supply or the acid-related release of toxic mineral ions. A laboratory experiment would be unable to mimic the complex interaction of abiotic and biotic factors in the real situation of the lake.

Nevertheless, it is only by searching for correlations and investigating them further that biologists can increase their understanding of ecology.

Essential Notes

A correlation may be either positive or negative. When one factor increases as another increases it is a positive correlation; when one increases while another decreases it is a negative correlation.

Fig H3
A line graph is a simple way of showing a correlation between two variables. You should be able to look at a graph and describe it in one or two sentences.

a Positive correlation: as x increases, y increases

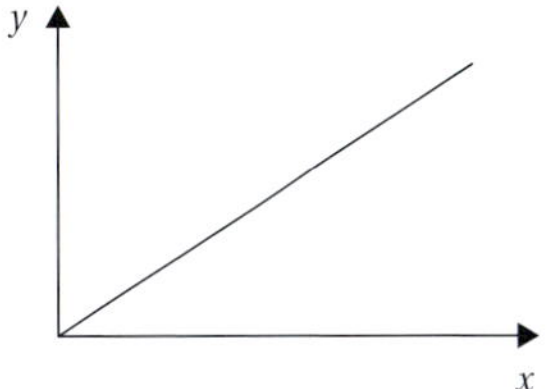

b Negative correlation: as x increases, y decreases

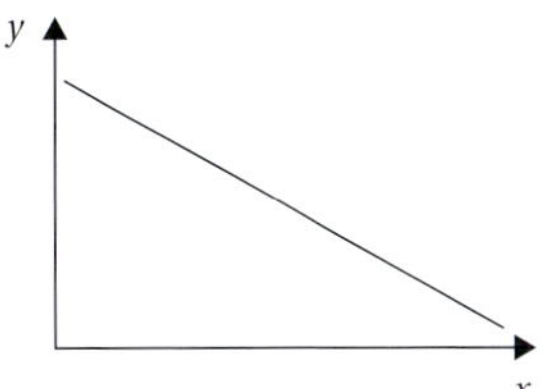

c As x increases, y increases up to a point, after which increasing x has no effect

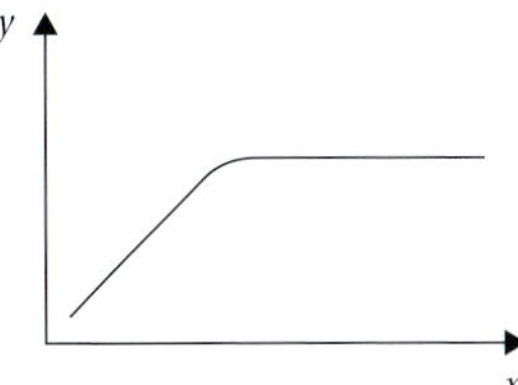

Practice exam-style questions

1 A group of students set out to investigate the distribution of seaweeds and animals on a rocky shore. They carried out a continuous belt transect along an exposed and a sheltered rocky shore and mapped out their findings (Fig E1).

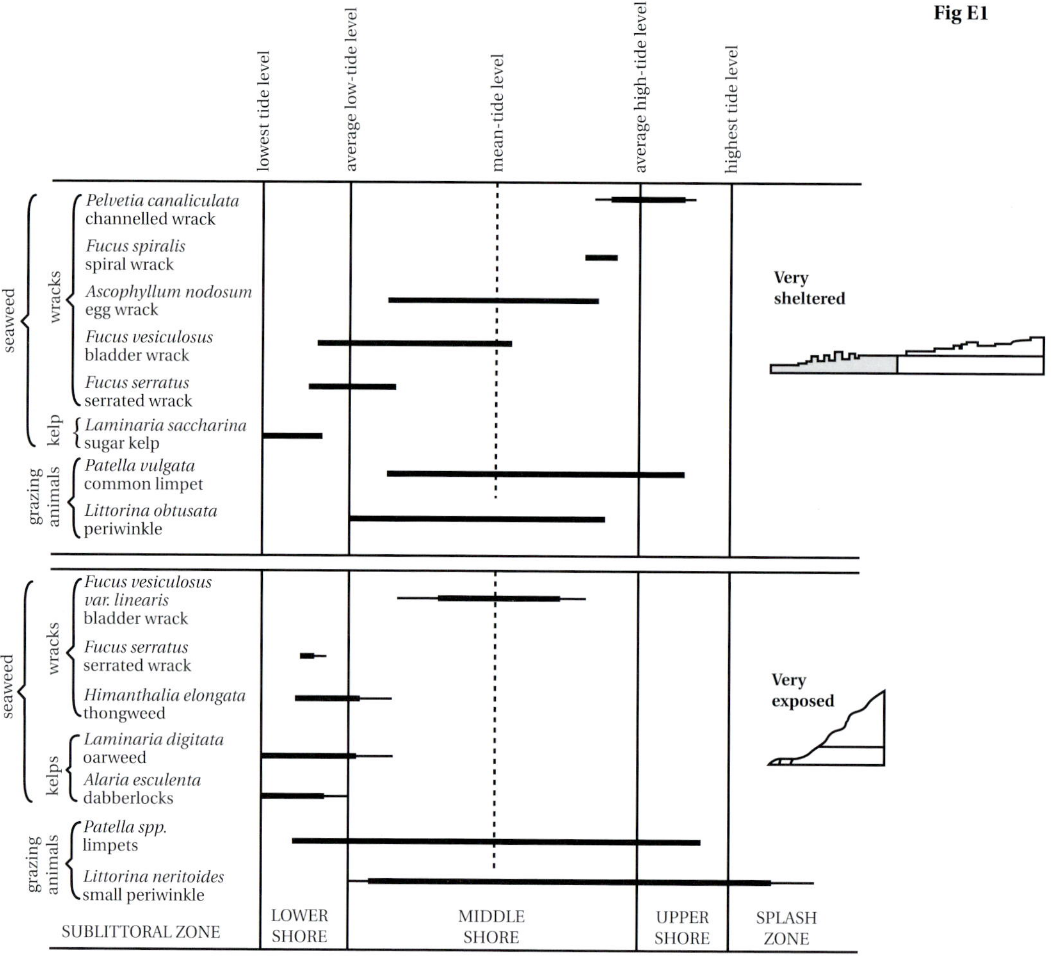

(a) Give one ethical and one safety consideration that would need to be taken into account before starting such an investigation.

____

____ 2 marks

(b) Explain how a continuous belt transect is carried out.

____

____

____ 3 marks

(c) Explain why a belt transect was a more suitable choice of sampling method than throwing quadrats.

2 marks

(d) Does the transect give qualitative or quantitative data? Explain your answer.

2 marks

(e) (i) Name two species that are only found on the sheltered shore.

1 mark

(ii) Suggest two abiotic factors that could account for the distribution of these species.

2 marks

Total marks: 12

2 In an investigation into the population size of the woodmouse *Apodemus sylvaticus* (also called the long-tailed field mouse), Longworth mammal traps baited with hamster food were used to capture the mice. On the first night 56 mice were captured. On the second night 61 were captured, of which 12 were marked.

(a) (i) State two precautions that should be taken when marking the mice.

2 marks

(ii) Explain why the trap should be checked regularly.

2 marks

(iii) Using the Lincoln index:

$$\text{population} = \frac{M \times C}{R}$$

Where:

M = Total number of animals captured and marked on the first visit.

C = Total number of animals captured on the second visit.

R = Number of animals captured on the first visit that were then recaptured on the second visit (i.e. number in second sample that were marked).

Estimate the population size of the mice.

2 marks

The woodmice are part of the woodland food chain shown below (Fig E2).

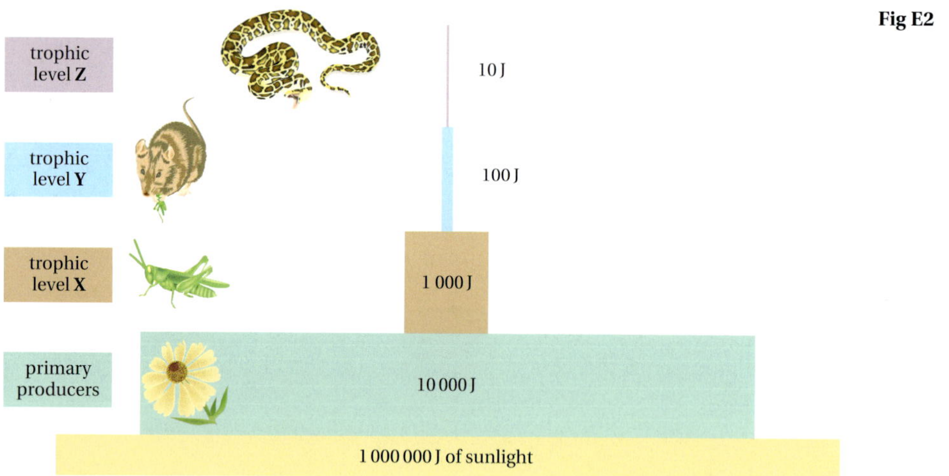

(b) Name trophic level X.

____ 1 mark

(c) Explain why so little of the energy in the producers is transferred to trophic level Y.

____

____

____ 3 marks

Total marks: 10

3 The diagram below (Fig E3) illustrates the energy conversion efficiency of one particular species of caterpillar.

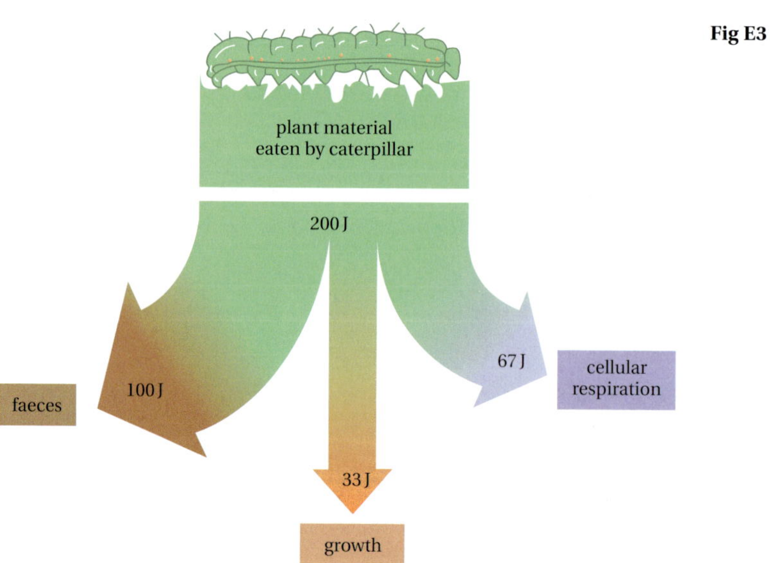

(a) Explain why such a lot of energy is lost in the faeces.

2 marks

(b) Suggest how the pattern of energy loss would be different if this herbivore was a mammal such as a rabbit. Explain your answer.

2 marks

(c) Explain how the carbon contained in the faeces could be used by a plant to make sugars.

4 marks

Total marks: 8

4 A group of students investigated the effect of grazing by rabbits. They wanted to know if grazing had an effect on species diversity. They compared two patches of land – one that was grazed by rabbits and one that was fenced off.

(a) Suggest a null hypothesis for this investigation.

1 mark

(b) Explain how you would place the quadrats without bias.

2 marks

Concentrating on the seven most abundant species in the area the students recorded the average number of individuals per quadrat, as shown in Table E1. They performed a statistical test to see if there was a significant difference in the distribution.

Table E1

Species	Grazed	Ungrazed	Significance (value of p)
Grass 1 (*Agrostis*)	25	21	0.5
Grass 2 (*Festuca*)	14	16	0.5
Buttercup	7	8	0.5
Dandelion	4	9	<0.05
Willow	0	11	<0.01
Bilberry	2	9	<0.01
Trefoil	0	4	<0.01

(c) What does the word ‘significant’ mean when applied to statistics?

1 mark

(d) Suggest which statistical test was used on these results.

1 mark

(e) What conclusions can be drawn from the results?

3 marks

(f) If the investigation were carried on for the next few decades, describe the changes that would occur in the fenced off (ungrazed) patch of land.

5 marks

Total marks: 13

5 *Paramecia* are single-celled protoctists found in ponds and other waterways. They feed on algae and other microscopic organisms. In a classic experiment, two species of *Paramecia* were grown both separately and together. The graphs (Fig E4) show the population growth curves for both species in both investigations.

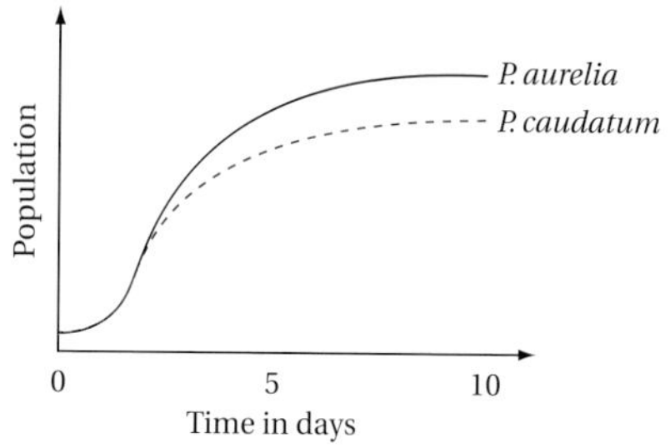

Fig E4

Graph 1 – grown separately

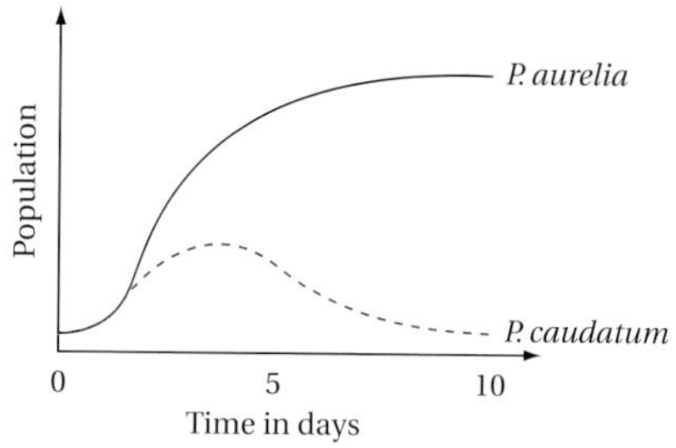

Graph 2 – grown together

(a) Suggest why growth is slow for the first day or so.

2 marks

(b) Account for the rapid growth of both species between days 1 and 4 when grown separately.

3 marks

(c) Suggest an explanation for the pattern of population growth in Graph 2.

2 marks

Total marks: 7

6 The graph (Fig E5) shows the United Nation's population projections by location up to the year 2050.

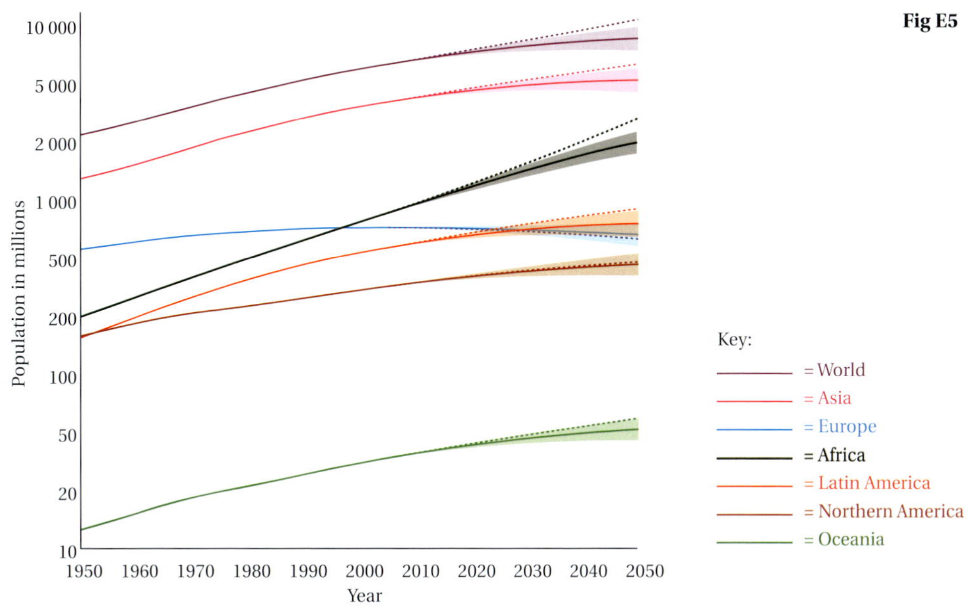

(a) What was the population of Africa in 1980?

1 mark

(b) Suggest two reasons for the rapid increase in the African population since 1950.

2 marks

(c) Suggest two reasons for the decline in the European population since 1990.

2 marks

(d) Suggest why there is a shaded area above and below the lines after about 2010.

1 mark

(e) Suggest two reasons why it is difficult to predict future population growth.

2 marks

Total marks: 8

7 Table E2 shows changes in fertility rate and the mean age of the mother at birth for the UK, 1981 to 2001. The fertility rate refers to the mean number of children per mother.

Table E2

	Fertility rate	Mean age of mother at birth
1981	1.82	26.8
1986	1.78	27.0
1991	1.82	27.7
1996	1.73	28.6
2001	1.63	29.2

(a) Describe the patterns shown in these results.

____

____ 2 marks

(b) Suggest reasons for each pattern identified in (a).

____

____ 2 marks

(c) What do the data suggest about the stage of demographic transition in the UK?

____ 1 mark

(d) Suggest two other factors that could be contributing to the population structure of the UK.

____

____ 2 marks

Total marks: 7

8 The graph (Fig E6) shows the population sizes of the snowshoe hare and the Canadian lynx over an 80-year period, taken from the records of Canadian fur trappers.

Fig E6

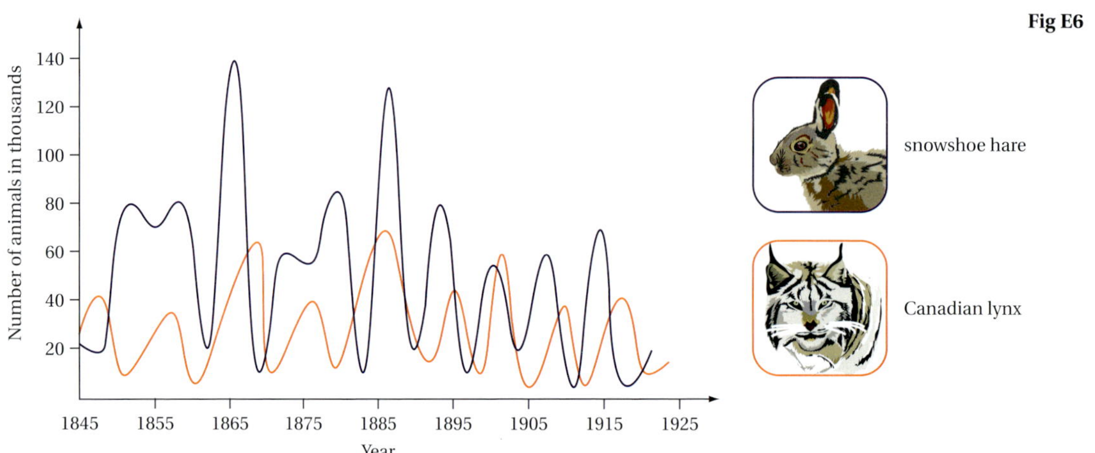

(a) What year saw the biggest fall in snowshoe hare numbers?

____ 1 mark

(b) Describe and explain the relationship between the two populations.

____

____

____

____ 4 marks

(c) Explain why relationships between predator and prey are rarely as clear as the one illustrated above.

____

____ 2 marks

Total marks: 7

9 The diagram below (Fig E7) shows a transect across a sand dune.

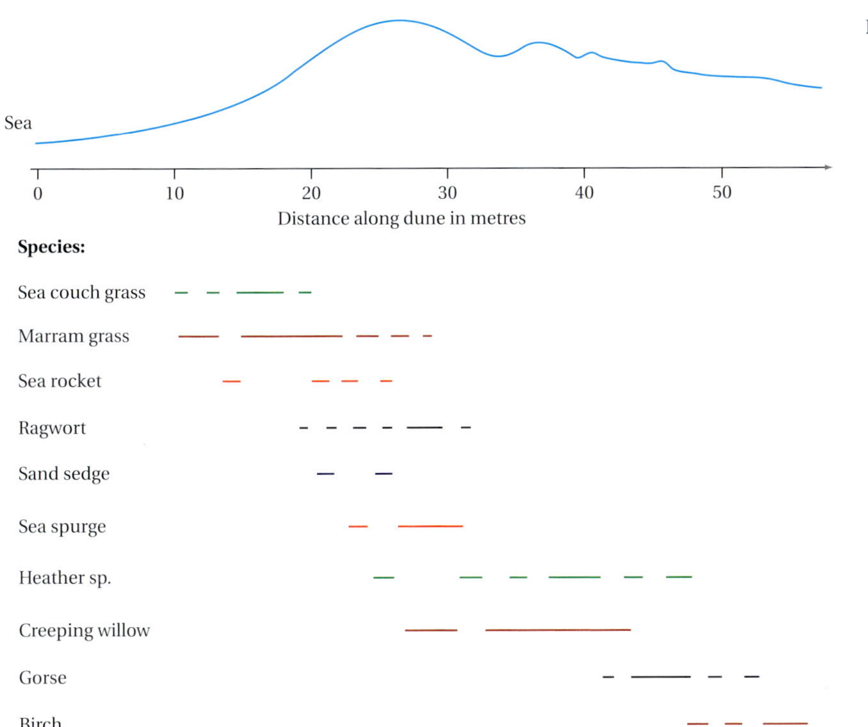

Fig E7

(a) Suggest and explain why an interrupted belt transect was used rather than a continuous belt transect.

2 marks

(b) List three abiotic factors that make it difficult for plants to grow in sand dunes.

3 marks

(c) Name the pioneer species shown in the diagram.

1 mark

(d) Suggest two adaptations that allow these pioneer species to grow in sand.

2 marks

Total marks: 8

10 (a) Cattle spread generous amounts of dung (faeces) onto grassland. Explain how the nitrogen in the protein in the dung becomes available to the grass.

6 marks

(b) Outline the process of eutrophication.

5 marks

Total marks: 11

11 The diagram below (Fig E8) shows a chloroplast.

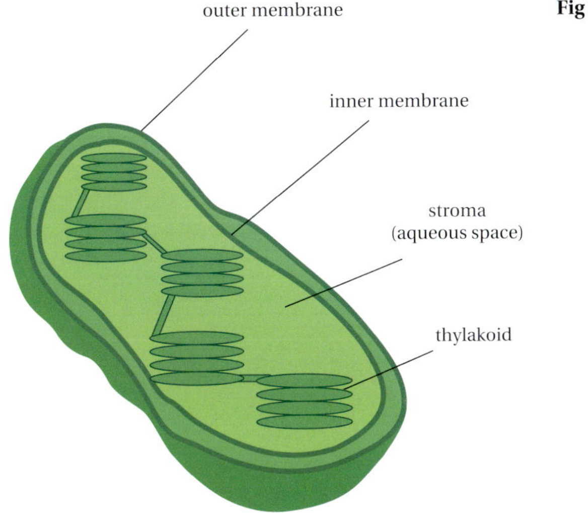

Fig E8

(a) Which one of the structures shown in the diagram:

(i) ... is the site of the light-dependent reactions?

______________________________ 1 mark

(ii) ... is the site of the light-independent reactions?

______________________________ 1 mark

The graph (Fig E9) shows the effect of light intensity on the rate of photosynthesis.

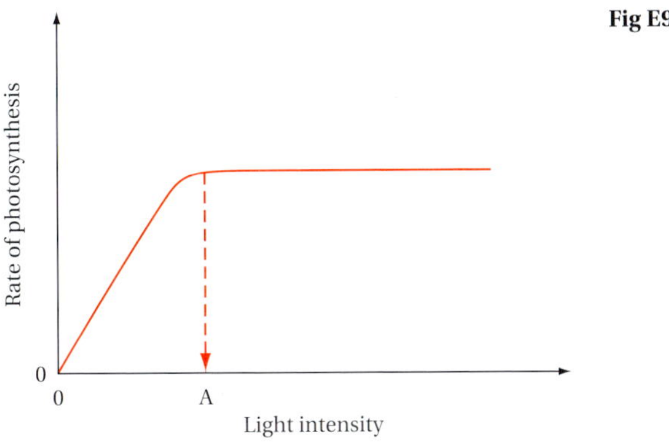

Fig E9

(b) Describe and explain the shape of the curve.

____

____

____ 3 marks

Total marks: 5

12 Match each word with its correct definition.

(a) Glycolysis	**(i)** Produces sugar from CO_2
(b) Chlorophyll	**(ii)** Energy currency of the cell
(c) Respiration	**(iii)** Reduced co-enzyme, electron-carrying molecule
(d) Calvin cycle	**(iv)** Breakdown of glucose to pyruvate
(e) NADH	**(v)** Produces ATP, NADPH and oxygen
(f) Krebs cycle	**(vi)** Central 5-carbon compound in Calvin cycle
(g) Electron transport chain	**(vii)** Takes place on inner mitochondrial membrane
(h) Light-dependent reaction	**(viii)** Converts a derivative of pyruvate to CO_2
(i) ATP	**(ix)** Emits excited electrons
(j) RuBP	**(x)** Oxidation of sugars to produce ATP

Total marks: 10

13 Albinism in humans is caused by an autosomal recessive allele a. The dominant allele A gives normal colouration. A normal (non-albino) couple have three children: two are normal, one is albino.

(a) What were the genotypes of the parents?

____ 1 mark

(b) What is the probability that their next child will be an albino?

____ 1 mark

(c) One of the normal children who carries the albino allele has a child with a partner who also has normal colouration. What predictions can be made about the colouration of their children? Explain your answer.

____

____

____ 2 marks

(d) The albino child has a child with a partner of normal colouration. What predictions can be made about the colouration of their children? Explain your answer.

____

____

____ 2 marks

Total marks: 6

14 One type of colour blindness is caused by a single gene located on the X chromosome. Normal individuals possess the allele C, which is dominant over the allele c which causes colour blindness. A female who carried the disease would have the genotype X^CX^c.

The pedigree diagram (Fig E10) shows three generations of the same family.

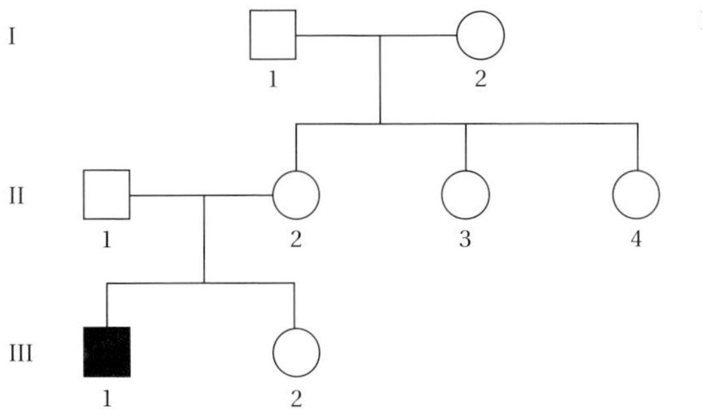

Fig E10

(a) Give the genotype of individuals 1 and 2 in each of the three generations. Some individuals may have more than one possible genotype.

____

____

______________________________ 3 marks

(b) The affected individual in generation 3 has a partner who has normal vision but is a carrier. What are the possible genotypes and phenotypes of their children? Show your working.

____

____

______________________________ 3 marks

Total marks: 6

15 In guinea pigs, black coat (B) is dominant to brown coat (b) and short hair (H) is dominant to long hair (h). These genes are autosomal and not linked.

(a) Explain what is meant by:

(i) Autosomal

______________________________ 1 mark

(ii) Not linked

______________________________ 1 mark

A guinea pig breeder has a pure-breeding, long-haired, brown male and a pure-breeding, short-haired, black female. A pet shop wants a supply of long-haired, black guinea pigs.

(b) Explain what the breeder will have to do to ensure a supply of pure-breeding, long-haired, black guinea pigs.

5 marks

Total marks: 7

16 Sickle-cell anaemia is a genetic disease in which abnormal haemoglobin causes sickle ('banana')-shaped red blood cells.

Normal homozygous individuals (SS) have normal blood cells that are easily infected with the malarial parasite. Thus, many of these individuals become very ill from the parasite and may die. Individuals who are homozygous for the sickle-cell trait (ss) have red blood cells that readily collapse into sickle shapes when deoxygenated. Although malaria cannot grow in these red blood cells, individuals often die because of the genetic defect. However, individuals with the heterozygous condition (Ss) have some sickling of red blood cells, but generally not enough to be fatal. In addition, malaria cannot reproduce within these 'partially defective' red blood cells. Thus, heterozygotes tend to survive better than either of the homozygous conditions.

If 9% of an African population is born with the severe form of sickle-cell anaemia (ss):

(a) Use the Hardy-Weinberg equation to calculate what percentage of the population is heterozygous (Ss) for the sickle-cell gene.

4 marks

(b) Explain what would happen to the s allele in a population where malaria was eradicated.

2 marks

Total marks: 6

Answers

Question	Answer	Marks
1 (a)	*Ethical – any one from*: Minimise damage to the organisms/ecosystem. Don't leave litter/rubbish. *Safety – any one from:* Wear suitable footwear. Avoid steep/slippery/sharp rocks. Avoid rough seas/deep water.	2, one for each
1 (b)	*Any three from*: Place line from low water line to splash zone (or words to that effect). Place transect on line. Record presence/absence or percentage cover of organisms. Move transect along/repeat at regular intervals.	3, one for each
1 (c)	A transect shows gradual change. Quadrats are better for comparing two different areas.	2
1 (d)	The transect gives qualitative data. Transect shows what species are present, *not* how many/relative abundance.	2
1 (e) (i) **1 (e) (ii)**	*Any two from*: channelled, spiral, egg or bladder wrack; sugar kelp. *Any two from*: wave action; time exposed to air/desiccation (drying out); temperature; salinity/water potential (refers to evaporation from rock pools).	1, half for each 2, one for each
		Total 12
2 (a) (i) **2 (a) (ii)** **2 (a) (iii)**	Use gloves to avoid being bitten. Mark mouse so as not to cause pain/affect mobility/affect visibility to predators. Check the trap regularly so that trapped animals ... do not starve to death. Population size of the mice: $\frac{(61 \times 56)}{12}$ = 285 (rounded to nearest whole number)	2 2 2
2 (b)	Trophic level X = primary consumers.	1
2 (c)	*Any three from*: A lot of plant material is indigestible. Animals cannot digest cellulose. Animals can only use the energy they absorb. A lot of energy is lost in faeces. A lot of energy is lost as heat to the environment/by-product of respiration. Mice are warm blooded/have a high rate of respiration.	3, one for each
		Total 10
3 (a)	A large proportion of plant material/cellulose is indigestible. Only the food that is digested and absorbed is available to the caterpillar.	2
3 (b)	*Any two from*: more lost as heat (by-product of respiration); warm blooded/use energy to maintain body temperature; so less energy for new growth.	2, one for each

Question	Answer	Marks
3 (c)	*Any four from*: Carbon can be found in the organic compounds in the faeces. Carbon can be broken down/hydrolysed/digested by enzymes. It can be released from bacteria/fungi/decomposers in extracellular digestion. Soluble compounds are absorbed into the bodies of decomposers. Carbon dioxide is released into the atmosphere by respiration of decomposers. Carbon dioxide is absorbed by plants; fixed into sugars by photosynthesis.	4, one for each
		Total 8
4 (a)	Grazing will have no effect on species diversity.	1
4 (b)	*Any two from*: Map out the area/make grid/coordinates. Select coordinates at random. Use table of random values/computer program.	2, one for each
4 (c)	It is unlikely that they occurred by chance.	1
4 (d)	Chi squared.	1
4 (e)	Grazing lowers species diversity. Grazing has no effect on the grasses or buttercup. Grazing has a significant effect on dandelion/willow/bilberry/trefoil.	3, one for each
4 (f)	Increase in humus/nutrient content of soil; succession by more perennial/ woody plants; greater diversity of plant and animal species; climax community of deciduous woodland; named dominant species, e.g. oak, beech, birch.	5
		Total 13
5 (a)	*Any two from*: Organisms take time to ingest food; they increase in volume/grow; they need to become mature enough to reproduce; time needed to activate genes; time taken to synthesize enzymes.	2, one for each
5 (b)	Growth was rapid in both species because there were *no limiting factors*. *Any two from*: plentiful food; no predation; no waste accumulation.	3
5 (c)	*P. aurelia* has *out-competed P. caudatum.* *Possible explanations*: *P. aurelia* better at catching food –or– *P. aurelis* is producing a substance/toxin that inhibits *P. caudatum*.	2
		Total 7
6 (a)	500 000 000 (or 500 million).	1
6 (b)	More people are living beyond middle age/into old age. Due to better sanitation, vaccination, education, nutrition.	2
6 (c)	*Any two from*: Low birth rate. Couples have fewer than two children. Couples wait longer to have children. Standard of living/career is more important than having large family. Possibly, emigration.	2, one for each

Question	Answer	Marks
6 (d)	Uncertain nature of prediction.	1
6 (e)	*Any two from*: inaccurate census techniques; climate change; socio-economic advances; medical advances.	2, one for each
		Total 8
7 (a)	Fertlity rate is declining. Mean age of the mother is increasing.	2
7 (b)	*Any two from*: Women are spending longer being educated. Women want a higher standard of living before starting a family. Women want a career as well as a family. There is more readily available/effective contraception. Older women are less fertile.	2, one for each
7 (c)	*In stage 4*: contracting or low/stationary.	1
7 (d)	Immigration/emigration; change in life expectancy.	2
		Total 7
8 (a)	The biggest fall in snowshoe hare numbers was around 1868/69.	1
8 (b)	*Any four from*: The higher the hare population, the greater the population of lynx that can be supported (*or vice-versa*). There is a *carrying capacity* – a certain population of lynxes can be supported by a certain population of hares. The more hares there are, the more food for the lynx and the more cubs they can rear. The lynx population is linked to the hare population. An increase in hare numbers is followed by an increase in lynx numbers ... ... after a lag/time delay.	4, one for each
8 (c)	Most predators eat more than one prey species. So interactions between populations are more complex.	2
		Total 7
9 (a)	Long distances are involved. A continuous transect would take too long.	2
9 (b)	*Any three from*: very little water; very few nutrients/minerals/humus; constantly shifting sand; plants can get buried; high salt content; high wind causes desiccation (drying out).	3
9 (c)	Marram grass and sea couch grass (*accept both or either one*).	1
9 (d)	*Any two from*: extensive roots consolidate sand; deep roots reach water; rolled leaves/sunken stomata/hairs reduce transpiration.	2, one for each
		Total 8
10 (a)	*Any six from*: dung/faecal matter broken down by saprobiotic (saprophytic) decay; bacteria and fungi synthesise and secrete enzymes; proteins hydrolysed into amino acids; process of ammonification; amino acids broken down to release ammonium salts; process of nitrification; changes ammonium into nitrite; the nitrite into nitrate; nitrate is soluble; can be absorbed by plants.	6, one for each

Question	Answer	Marks
10 (b)	*Any five from*: waterways become over-fertile; due to leaching of minerals (nitrate, phosphate, etc.) from agricultural land; causes algal bloom; algae die/decay; high bacterial population uses most of the available oxygen; normal clean water/high oxygen organisms cannot survive; high BOD (biochemical oxygen demand) value.	5, one for each
		Total 11
11 (a) (i)	Thylakoid is site of light-dependent reactions.	1
11 (a) (ii)	Stroma is site of light-independent reactions.	1
11 (b)	*Any three from*: As light intensity increases so does rate of photosynthesis. Light is the limiting factor. At A some other factor is limiting. For example, carbon dioxide levels may be limiting.	3, one for each
		Total 5
12 (a)	(iv)	1
12 (b)	(ix)	1
12 (c)	(x)	1
12 (d)	(i)	1
12 (e)	(iii)	1
12 (f)	(viii)	1
12 (g)	(vii)	1
12 (h)	(v)	1
12 (i)	(ii)	1
12 (j)	(vi)	1
		Total 10
13 (a)	Both parents must be genotype Aa.	1
13 (b)	The probability is: 0.25, 25% or 1 in 4 (all mean the same).	1
13 (c)	If partner is AA, there is no chance that the children will be albino. If partner is also a carrier (i.e. Aa), there is a 1 in 4 chance.	2
13 (d)	If the new partner is AA, then all the children will have normal colouration. The children will be genotype Aa. If partner is Aa, there is a 1 in 2 (50%) chance that the children will be albino.	2
		Total 6
14 (a)	*1 mark for each generation all correct*: Generation 1 – Male = X^CY; Female = X^CX^c Generation 2 – Male = X^CY; Female = X^CX^c Generation 3 – Male = X^cY (i.e. colour blind); Female = X^CX^c or X^CX^C.	3

Question	Answer	Marks
14 (b)	*1 mark each (for parents, gametes and offspring)*: Parental genotypes - Male = X^cY; Female X^CX^c Gametes X^c, Y, X^C Offspring = X^CX^c (female, normal vision) X^CY (male, normal vision) X^cY (male, colour blind)	3
		Total 6
15 (a) (i)	Autosomal means: genes not located on sex chromosomes.	1
15 (a) (ii)	Not linked means: genes not located on same chromosomes.	1
15 (b)	*Any five from*: First generation are all Bbh – Hh. Second generation – all possible genotypes. The breeder wants BBhh individuals. Long-haired, black guinea pigs might be Bbhh or BBhh. To find out which, do a test cross with a white-coated individual. If all offspring are all black, individual was BBhh – pure breeding. If half black, half white, individual was Bbhh – not pure breeding.	(1 mark given for correct Punnett square) 5, one for each
		Total 7
16 (a)	The frequency of ss (i.e. q^2) is 0.09. Therefore the frequency of the q allele is 0.3. $p + q = 1$; therefore the frequency of the p allele is 0.7. The frequency of the Ss genotype is $2pq$, i.e. $2 \times 0.7 \times 0.3 = 0.42$ or 42%.	4
16 (b)	*Any two from*: Ss individuals no longer have an advantage. SS individuals would have an advantage over Ss and ss. Frequency of the s allele would decrease.	2, one for each
		Total 6

Glossary

Abiotic factor	Non-living environmental factor, such as temperature, pH, humidity, carbon dioxide level. Compare with *biotic factor*.
Acetate	A 2-carbon molecule produced by the *link reaction* (the oxidation of pyruvate) in aerobic respiration. Combines with *co-enzyme A* to form *acetyl co-enzyme A*. Acetate is basically what is left of the glucose after glycolysis and the *link reaction*.
Acetyl co-enzyme A (acetyl CoA)	Compound that enters the *Krebs cycle*. Co-enzyme A is a carrier molecule that picks up the acetate from the link reaction and delivers it to the Krebs cycle.
Active transport	Transport across a membrane against a concentration gradient. Requires energy from the cell (ATP).
Aerobic respiration	Respiration that uses oxygen. Results in the complete oxidation of the substrate which produces a lot more ATP than anaerobic respiration. Takes place in the mitochondria. Compare with *anaerobic respiration*.
Allele	Alternative form of a gene. For example, a flower colour gene could have two alleles, R and r. Allele R codes for red flowers, r for white. Some genes have no alleles, others have many.
Allele frequency	The frequency of an allele in a population. For example, if there are two alleles - R and r - and the frequency of the R allele is 0.85 (85%), then the frequency of the r allele must be 0.15 (or 15%). Evolution can be defined as a change in allele frequency.
Allopatric speciation	Type of speciation in which a new species develops when physically separated from the original population. Compare with *sympatric speciation*.
Ammonification	Vital step in the nitrogen cycle. Process of decay in which proteins and amino acids are broken down by bacteria and fungi to release ammonium salts. Also called *saprobiotic nutrition*. Usually followed by nitrification in the nitrogen cycle.
Anaerobic respiration	Respiration without oxygen. Incomplete oxidation of the substrate (usually glucose) yields a small amount of ATP and an intermediate substance - pyruvate - that is converted into lactate or carbon dioxide/ethanol, depending on the organism. Compare with *aerobic respiration*.
Anthropogenic	Of human origin. Some atmospheric carbon dioxide is anthropogenic, but some is there due to natural causes.
ATP	Adenosine triphosphate. Compound that acts as immediate energy source for metabolic reactions. *Respiration* makes ATP from ADP and phosphate; many other processes use ATP, such as muscular contraction, active transport.
Autosome	A chromosome other than a sex chromosome. Humans have 22 pairs of autosomes and one pair of sex chromosomes.
Autotroph	Organism that makes its own food using an external energy source - usually sunlight - and a simple inorganic supply of carbon, usually carbon dioxide. All producers are autotrophs. See *chemoautotroph and photoautotroph*.
Average life expectancy	Statistical measure of the expected life span of an individual in a population. Worked out using a *survival curve*: average life expectancy is the age at which 50% of the individuals in the sample are still alive.

Biodiversity	Biological diversity; a measure of the number of species living in a certain area. Some habitats have a very high biodiversity (for example, rainforests, coral reef) and some a very low biodiversity (for example, polar regions and deserts). Reduction in biodiversity due to human activity is a major concern.
Biomass	The mass of living material. Usually applies to the mass of a particular population or trophic level, for example, the mass of all the grass in an ecosystem, but can apply to mass per unit area (such as m^2).
Biotic factor	Environmental factor caused by other organisms, such as food supply, *predation* or disease. Contrast with *abiotic factor*.
Calvin cycle	Series of reactions that make up the *light-independent reaction* of photosynthesis. Essentially involves the use of ATP and NADPH to reduce carbon dioxide into glucose.
Carbon cycle	Global cycle of reactions in which carbon dioxide is removed from the atmosphere by photosynthesis and returned by respiration and combustion. Human activity is shifting the balance.
Carrying capacity	The size of population that can be supported by a particular ecosystem. Often limited by food supply and predation.
Chemoautotroph	Organism that makes its own organic compounds using energy from chemical reactions (usually oxidation). These are all bacteria and nitrifying bacteria are common examples. Compare with *photoautotrophs*.
Chi squared (χ^2)	Statistical test that compares observed values (i.e. those gathered in the investigation) with those you would expect if there were no correlation. Named after the Greek letter chi, χ.
Chlorophyll	Green pigment with a central role in photosynthesis. There are several different pigments that absorb different wavelengths of light, but all essentially absorb light and emit high-energy electrons.
Chloroplast	Organelle of photosynthesis. Contains chlorophyll molecules housed on flat discs (*thylakoids*) piled up into a *granum*.
Chromosome	Highly condensed ('super-coiled') DNA molecule that appears in a cell just before cell division. The name means 'coloured body'.
Climax community	Situation where an ecosystem has matured and stabilised after a period of *succession*; for example, deciduous woodland, tropical rainforest. Characterised by a few dominant species, usually trees.
Codominance	A situation where neither allele is recessive and so if both are present in the genotype, both are expressed in the phenotype. Seen in type AB blood.
Co-enzyme A (CoA)	In aerobic respiration, compound that combines with *acetate* to form *acetyl co-enzyme A* (*CoA*). CoA combines with the acetate formed in the *link reaction* and transfers it into the *Krebs cycle*.
Colonisation	In the development of ecosystems, colonisers are the first organisms to take hold in a non-living environment; for example, lichens on bare rock. Colonisers improve the abiotic environment so that a wider variety of organisms can live there, and succession begins.

Combustion	Formal word for 'burning' – the complete oxidation of substances at high temperature. If enough oxygen is available, combustion of organic material produces carbon dioxide and other gases.
Community	The living component of an ecosystem; i.e. the interacting populations of all the different species.
Competitive exclusion principle	Idea stating that no two species can occupy the same niche in an ecosystem.
Consumer	Organism that cannot make its own organic molecules, and so must obtain them ready-made from other organisms. Animals, fungi and most bacteria are consumers. Contrast with *producer*.
Continuous belt transect	Sampling method used to show the change in species from one area to another. In a continuous belt transect, quadrats are placed alongside each other with no gaps.
Cristae	Folds in the inner membrane of a mitochondrion; site of the electron transport system.
Cytosol	The fluid part of the cytoplasm; the liquid between the organelles.
Deciduous forest	Forest in which most of the dominant tree species shed their leaves; for example, oak, ash, beech; the *climax community* in the UK.
Deforestation	Chopping down trees to sell the wood or to make way for agriculture, or both.
Deme	An interbreeding population.
Demographic transition	Change in population structure, from a country or population with high birth rate and high death rate, through one with lower death rate (and therefore an expanding population) to one with low birth rate and low death rate.
Denitrification	In the nitrogen cycle, a bacterial process that converts soluble nitrate into nitrogen gas (N_2), thereby losing it from the ecosystem. Tends to occur in anaerobic conditions, such as waterlogged soil.
Denitrifying bacteria	Bacteria responsible for *denitrification*.
Density-dependent factor	Environment factor that depends on the size or density of a population. For example, the higher the population, the greater the competition for food.
Detritivore	An animal that feeds off detritus, such as the earthworm and woodlice. Detritivores should not be confused with saprobionts: detritivores are animals with intestines that actually eat the detritus, while saprobionts have no intestines and feed by extracellular digestion.
Detritus	Rotting organic matter such as dead leaves, wood, animal bodies and faeces.
Dihybrid cross	Cross involving two separate genes. Notation used is usually something like: AABB × AaBb.
Diploid	Cell or organism that possesses two sets of chromosomes. Often written as $2n$. In humans, $2n = 46$.
Directional selection	Type of natural selection that favours one extreme of phenotype, for example tallest, quickest, heaviest.
Disruptive selection	Type of natural selection that favours both extremes of phenotype over the mid range.

Diversity	Measurement of the number of different species present in an ecosystem. The diversity of different ecosystems, or change over time, can be measured with a diversity index.
Dominant	An allele which, if present, is expressed in the phenotype. Compare with *recessive*.
Ecosystem	Natural unit such as a lake, woodland, coral reef, etc. that contains many different species together with the non-living components.
Electron transport system	The final stage of aerobic respiration in which electrons are transported along a series of proteins on the inner mitochondrial membrane. The resulting redox reactions release energy that is used to pump H^+ ions into the outer mitochondrial membrane. Diffusion of H^+ back into the matrix through ATPase enzymes powers ATP synthesis. Electron transport systems also play an important role in the light-dependent reaction of photosynthesis.
Environmental resistance	The inevitable restriction of population growth, which always slows down sooner or later due to limiting factors.
Ethanol	Ethyl alcohol, C_2H_5OH. It makes you drunk.
Eutrophication	In ecology, a situation where a waterway is over-fertile due to excess fertiliser. Main stages: Algal bloom → bacterial bloom → oxygen shortage.
Exponential phase	In population growth, stage of rapid growth due to a lack of limiting factors, i.e. when conditions are favourable. See also *logarithmic phase.*
Extracellular digestion	Mode of nutrition in which organisms (bacteria or fungi) synthesise and release enzymes that digest the surrounding organic material. The organisms then absorb the soluble products. Basically, this is why things rot.
Factor VIII	Protein involved in blood clotting. Inability to make factor VIII is the cause of one type of *haemophilia*.
Gene	A length of DNA that codes for the production of a particular polypeptide or protein.
Gene pool	The sum total of the *alleles* circulating in an interbreeding population, or deme.
Genotype	The particular alleles an organism possesses.
Glycerate 3-phosphate (GP or G3P)	A 3-carbon compound, it is the first sugar made in the light-independent reaction of photosynthesis, sometimes referred to as GALP – glyceraldehyde phosphate.
Glycolysis	The first stage of respiration, in which glucose is converted into *pyruvate*. It's a universal process, taking place in the cytoplasm of all cells in all organisms, whether anaerobic or aerobic. Makes 2 ATP and 2 NADH molecules per glucose molecule.
Granum	A pile of *thylakoids* in a *chloroplast*. Site of chlorophyll and therefore of the light-dependent reaction in photosynthesis.
Gross primary production (GPP)	The rate at which the producers in an ecosystem make biomass (organic material such as sugars, starches and cellulose) during photosynthesis. See also *net primary production* (*NPP*).
Habitat	The external environment in which an organism lives.
Haemophilia	Sex-linked genetic disease in which blood fails to clot properly due to an inability to make *factor VIII*. Caused by a recessive, sex-linked allele. Males can't be carriers, so if they inherit the allele, they have the disease.

Hardy-Weinberg	Law which states that in a large, randomly mating population, allele frequencies will remain constant from one generation to the next, unless there is mutation, immigration, emigration or selection.
Heterotroph	An organism that cannot make its own food, so must obtain food molecules ready-made from its surroundings. All consumers are heterotrophs – animals, fungi, most bacteria and many protoctistans.
Heterozygous	When an organism has two different alleles of a particular gene. Usually written as Aa, Bb, etc.
Homozygous	When an organism has two of the same alleles. Shown as AA or BB.
Humus	Rotting organic matter in soil – a sticky mass of dead plants, dead animals and faeces. Humus is vital to soil because it improves texture and allows it to hold more water. Saprobionts act on humus to release the nutrients – such as nitrate and phosphate – that plants need to grow.
Hydrolysis	Literally 'splitting using water'. Reactions in which larger molecules are split into smaller molecules by the addition of water. Most digestion occurs by hydrolysis.
Incomplete dominance	In genetics, a situation where the heterozygote shows an intermediate phenotype; for example, flower colour in carnations – allele R gives red flowers; allele r codes for white flowers; heterozygote (Rr) flowers are pink.
Interrupted belt transect	Sampling method used to show the change in species from one area to another. In an interrupted belt, quadrats are placed at intervals along the transect. Suitable for a longer transect where the change is gradual. Compare with *continuous belt transect.*
Interspecific competition	Competition between individuals of *different* species.
Intraspecific competition	Competition between individuals of the *same* species.
Isolation	Vital step in speciation. Two populations become isolated when they cannot interbreed. Natural selection will then act in different ways on the isolated populations.
Krebs cycle	Stage in aerobic respiration that takes place in the matrix of a mitochondrion. Begins with *acetyl co-enzyme A* and produces reduced co-enzymes (NADH and $FADH_2$), ATP, and CO_2 as a by-product.
Lactate	A 3-carbon molecule produced in *anaerobic respiration* in various organisms. Builds up during muscle fatigue. Lactate is lactic acid in solution.
Lag phase	In population growth, an initial period of slow growth.
Leaching	When mineral ions (nitrate, phosphate, etc.) are washed out of soil, such as when rainfall washes ions away from fertilised farmland and into waterways, leading to eutrophication.
Legume	Plant that has root nodules containing *nitrogen-fixing bacteria,* which convert N_2 gas into nitrate. Includes peas, beans, lentils, clover and peanuts.
Light-dependent reaction	First stage of photosynthesis – chlorophyll absorbs light and emits two high-energy 'excited' electrons. ATP and NADPH are the vital products of the light reaction. Takes place on the thylakoids.
Light-independent reaction	Second stage of photosynthesis in which carbon dioxide is used to make sugar, via the reactions of the *Calvin cycle.* Happens in the *stroma* (fluid) of chloroplasts.

Limiting factor	The factor that is in short supply. If supply is increased, the rate of the process increases too. Light is a common limiting factor for photosynthesis.
Lincoln index	Formula for estimating population size using the mark-release-recapture method.
Link reaction	In respiration, stage that links glycolysis to the Krebs cycle. Pyruvate is converted into acetate, which is picked up by *co-enzyme A* to become *acetyl co-enzyme A*. Also known as pyruvate oxidation.
Linkage	When two genes appear on the same chromosome they are said to be linked. In this case they will not be separated by independent assortment.
Locus	Position of a gene/allele on a chromosome.
Logarithmic phase	In population growth, period of rapid growth because there are no limiting factors. Also known as *exponential phase.*
Mark-release-recapture	Method for estimating the population of a particular animal species. The animal is trapped, harmlessly marked and then released. On a second occasion, the population size can be estimated from the number in the second sample that are already marked. See *Lincoln index.*
Matrix	Fluid in the centre of mitochondria. Site of the link reaction and the *Krebs cycle.*
Meiosis	Type of cell division that produces four daughter cells, all haploid and genetically unique.
Mendelian population	An interbreeding population. Also called a *deme.*
Microhabitat	The environmental conditions in an organism's immediate surroundings; for example, the conditions under a stone on a river bed will be completely different from those above it – light, temperature, oxygen levels, flow rate and risk of predation will all be different.
Mitochondrion	Organelle found in most eukaryote cells. Site of *aerobic respiration.*
Mitosis	Type of cell division in which one cell divides to produce two identical daughter cells. Multicellular organisms grow and develop by mitosis.
Monoculture	Agricultural practice of growing large fields consisting of just one crop.
Monohybrid cross	A cross involving a single gene.
Multiple allele	An allele with more than two variants.
Mutation	A change in the genetic material of a cell or organism. There are gene mutations, in which the base sequence of the gene is altered, and chromosome mutations, in which whole blocks of genes are changed.
Mutualism	An association between two organisms where both species benefit. Old name: symbiosis. Examples include lichens (algae inside fungi), corals (algae inside animals) and legumes (nitrogen-fixing bacteria in plant roots).
NAD^+	Co-enzyme that plays a central part in respiration. NAD^+ picks up an electron to become NADH, reduced co-enzyme.
$NADP^+$	Co-enzyme that plays a central part in photosynthesis. $NADP^+$ picks up an electron to become NADPH, reduced co-enzyme.
Net primary production (NPP)	The rate of accumulation of biomass by photosynthesis after respiration has been taken into account. NPP = GPP – respiration.
Niche	Concept that describes an organism's position in an ecosystem.

Nitrification	Vital two-stage process in the nitrogen cycle. Ammonium salts are converted first into nitrite, then into nitrate.
Nitrifying bacteria	Bacteria that carry out nitrification. See *nitrogen cycle.*
Nitrogen cycle	Cycle of reactions that re-uses nitrogen. Plants absorb nitrate and use it to make proteins and nucleic acids. In this form nitrogen passes up the food chain until it is broken down into nitrate again by the action of saprobionts and *nitrifying bacteria.*
Nitrogen fixation	Process in which nitrogen gas is converted, via ammonia (NH_3), into soluble nitrate so that it is available to organisms. Can happen in the atmosphere during electrical storms but is more commonly done by bacteria living free in soil/water or inside plant roots (see *legumes*).
Nitrogen-fixing bacteria	Bacteria that carry out nitrogen fixation. See *Rhizobium.*
Nitrogenase	Enzyme found in nitrogen-fixing bacteria that converts nitrogen gas into ammonia, which can then be converted into nitrate.
Null hypothesis	A hypothesis that takes a negative position on an investigation, such as 'there is no link between cancer and smoking' so that it can be disproved/refuted by statistics. It is the nature of scientific investigation that you can never prove anything, but you can disprove things. By testing the null hypothesis with statistics, you can reject it and so gather support for your hypothesis. See *chi squared, Spearman Rank.*
Oxidative phosphorylation	ATP made by the electron transport chain. In respiration there are two basic methods of making ATP – the other being *substrate level phosphorylation.* Oxidative phosphorylation yields far more ATP per glucose molecule (over 30 compared with just four).
Phenotype	The observable features an organism possesses. Simply put: phenotype = genotype + environment. Contrast with *genotype.*
Phosphorylation	The addition of a phosphate group to a compound, as in the synthesis of ATP from ADP.
Photoautotroph	Organism that makes organic compounds using energy from sunlight; i.e. an organism that photosynthesises. Compare with *chemoautotroph.*
Photolysis	In photosynthesis, the splitting of water to provide replacement electrons for those lost by chlorophyll in the light-dependent reaction. Vitally, oxygen is a by-product.
Photophosphorylation	The production of ATP during the light-dependent reaction of photosynthesis.
Photosynthesis	Series of reactions in which sunlight energy is used to synthesise organic compounds such as sugars. Vital process which is virtually the only route of energy into most ecosystems. Makes all the food on the planet, removes carbon dioxide and produces oxygen.
Pioneer species	In the development of ecosystems, pioneer species are the first to establish themselves. They can usually tolerate harsh conditions and low nutrient levels. Examples include lichen on bare rock, or marram grass on sand dunes. Also called *colonisers.*
Point transect	*Transect* in which the individual species touching a particular point on the line (for example, every 10 cm) are recorded.

Population	Group of individuals of the same species that can interbreed.
Population pyramid	Diagram that illustrates population structure, showing numbers of males and females in progressive five-year segments. Can be used to illustrate the stage of demographic transition of a particular population or country.
Predation	When one organism (the predator) eats another (the prey). An important *biotic factor*. The numbers of predator and prey are often inter-dependent.
Producer	Organism that can photosynthesise. They produce organic compounds such as sugars from inorganic compounds, notably carbon dioxide and water. Plants, algae and photosynthetic bacteria are all producers.
Punnett square	In genetics, a grid for organising all the possible outcomes of a cross.
Pyramids of biomass, numbers and energy	Diagrams that show relative amounts of biomass, numbers or energy at each trophic level. Pyramids of energy are always true pyramids (wide base, narrower at top).
Pyruvate	A 3-carbon compound produced in glycolysis from the splitting of glucose. One glucose molecule produces two pyruvate molecules.
Quadrat	In ecology field work, a sampling method used to compare two different areas. Often a small, 0.5 m square frame.
Qualitative data	Data that show 'what' rather than 'how much'.
Quantitative data	Data that show 'how much' as well as 'what'.
Recessive	In genetics, an allele that is only expressed in the absence of the *dominant* allele.
Reduced NAD	Electron-carrying co-enzyme with a vital role in respiration. Co-enzyme, often written as NAD^+, picks up an electron to become reduced co-enzyme, often written as NADH.
Reproductive success	The ability of an organism to pass its alleles/allele combinations on to the next generation. Sometimes referred to as 'fitness' and often simply measured by the number of offspring.
Respiration	Process that releases energy in organic molecules and transfers it to *ATP* so the cell/organism has instant energy available. Universal process, one of the seven signs of life.
Rhizobium	Genus of *nitrogen-fixing bacteria* found in the root nodules of legumes. An example of a mutualism, both bacteria and plants benefit from this association. *Rhizobium* cannot fix nitrogen outside the plant.
Rubisco	Ribulose-1,5-bisphosphate carboxylase. Enzyme that catalyses the reaction between RuBP and carbon dioxide in the light-independent reaction (Calvin cycle) of photosynthesis. Thought to be the most abundant protein on Earth.
RuBP	Ribulose bisphosphate. Vital 5-carbon compound in the light-independent reaction in photosynthesis.
Saprobiont	Preferred name for *saprophyte*. A decomposer that breaks down dead organic matter by extracellular digestion. Mainly bacteria or fungi, they secrete enzymes and absorb the soluble products of digestion.
Saprobiotic nutrition	Preferred name for *saprophytic decay*, a process in which bacteria and fungi break down dead organic matter by extracellular digestion.
Saprophyte	Old name for *saprobiont*.

Saprophytic decay	See *saprobiotic nutrition.*
Sere	One of the stages of succession in the development of an ecosystem.
Significance	In statistics, a result is significant if it is unlikely to have occurred by chance. Most statistical tests give a probability value, *p*. The smaller the *p* value, the more significant the results are likely to be. The significance value can be 0.05, 0.01, or even 0.001. A commonly accepted significance value is 0.05, meaning that if the value of *p* is less than 0.05 (or 5%) then the null hypothesis can be rejected.
Spearman Rank	Statistical test that assesses the correlation between two variables. For example, in the hypothesis 'the heavier the individual, the greater the calorific content of their diet', you could take a group of individuals, ranked according to weight, and measure the calorific content of their diet. The Spearman Rank test could then be used to assess whether the two ranks were significantly correlated or not.
Speciation	The process of forming a new species from pre-existing species.
Stabilising selection	Type of natural selection in which middle values have a selective advantage.
Stroma	The fluid part of a chloroplast. Contains all the enzymes and intermediates for the *light-independent reaction* in photosynthesis.
Substrate level phosphorylation	ATP produced from glycolysis and the Krebs cycle, a total of four molecules per glucose. Contrast with ATP made from *oxidative phosphorylation*, which is ATP generated by the electron transport system (over 30 molecules in total).
Succession	The development sequence of an ecosystem, in which one set of plant species changes the abiotic environment so that a different set of species takes over. Sequence continues until a *climax community* is established.
Survival curve	Graph plotted using the actual lifespan of a sample of 10 000 individuals. Can be used to calculate *average life expectancy*.
Sustainable	In agriculture, food production that can be maintained year on year without a depletion of natural resources or permanent damage to the ecosystem.
Sympatric speciation	One type of speciation, where isolation takes place even though the two populations live together. Clear examples are rare. Compare with *allopatric speciation*.
Thylakoid	Flat disc-shaped membrane that houses the *chlorophyll* inside a *chloroplast*. A pile of thylakoids forms one *granum*.
Transect	In ecology field work, a sampling method in which the organisms present along a line are recorded to show change from one area to another.
Triose phosphate	A 3-carbon sugar that is the first carbohydrate produced in the light-independent reaction of photosynthesis (also known as the Calvin cycle).
Trophic level	A 'feeding' level in the food web. The first trophic level usually consists of producers, though detritivores can also support a food web. The next trophic level is the primary consumers (usually herbivores), then secondary consumers (carnivores) and sometimes tertiary consumers or even higher. Some animals are omnivorous, for example, rats, pigs and monkeys. They have a very varied diet and so can occupy more than one trophic level.
Wetlands	Areas where the soil is saturated with water, as in a swamp, bog, marsh, estuary, mangrove, tundra and flood plain. They are important habitats covering huge areas of land.

Index